T 29592 $29.95

ROBOTICS

NEW EDITION

SCIENCE & TECHNOLOGY IN FOCUS

ROBOTICS
N E W   E D I T I O N

Intelligent Machines for the New Century

Ellen Thro

Facts On File, Inc.

ROBOTICS: Intelligent Machines for the New Century, New Edition

Copyright © 2003, 1993 by Ellen Thro

Facts On File, Inc.
132 West 31st Street
New York NY 10001

Library of Congress Cataloging-in-Publication Data

Thro, Ellen.
Robotics / Ellen Thro.—New ed.
p. cm.—(Science and technology in focus)
ISBN 0-8160-4701-4 (hardcover)
1. Robotics—Popular works. I. Title. II. Series.
TJ211.15 .T457 2003
629.8'92—dc21   2002001233

Facts On File books are available at special discounts when purchased in bulk quantities for businesses, associations, institutions, or sales promotions. Please call our Special Sales Department in New York at 212/967-8800 or 800/322-8755.

You can find Facts On File on the World Wide Web at
http://www.factsonfile.com

Text design by Erika K. Arroyo
Cover design by Nora Wertz
Illustrations by Patricia Meschino

Printed in the United States of America

MP FOF 10 9 8 7 6 5 4 3 2 1

This book is printed on acid-free paper.

CONTENTS

ACKNOWLEDGMENTS

My gratitude goes to the many people who helped make this new edition of *Robotics* possible.

At Facts On File, Inc., Senior Editor Frank K. Darmstadt provided helpful suggestions at just the right moments. Thanks also to Cynthia Yazbek, Associate Editor; Michael G. Laraque, Chief Copy Editor; Faith K. Gabriel, Copy Editor; and to those who designed the interior and exterior of the book. Independent photo researcher Lisa Kirchner pulled rabbits out of hats to get the pictures I needed.

I also reiterate my thanks to the team that helped produce the first edition of *Robotics:* Barbara Lucas of Lucas-Evans Books, Commander Bart Everett of the Naval Oceans System Center, Pat Kelly at the Balboa Naval Hospital, and Nicole Bowen, editor of the original book series, and her staff.

INTRODUCTION

Miniaturization. Sensory feedback. Voice recognition.

Since the original edition of *Robotics* was published in 1993, advances in these areas have transformed the development and use of intelligent robotics. Advanced robots have already been taken into homes, offices, hospitals, and other ordinary environments.

For instance, life-sized robotic "infants," such as My Real Baby, are learning to interact with their environment. ASIMO, an advanced intelligent humanoid (humanlike) robot, can walk and interact with people. Another example is Aibo, the doglike and dog-sized robot now on sale as a family "pet," capable of learning to respond to commands and "playing."

On a grimmer note, the phrase "hazardous environment" expanded to include the World Trade Center site, where small portable robots for the first time helped human researchers and rescue dogs find victims' remains after the tragedy of September 11, 2001.

New sensory materials that are processed into "muscles" and "human fingers" are available for robots working in space environments. Robots are also being used in construction of the *International Space Station*.

The first robotic nanomachine now exists—a few billionths of a meter in size. Medical scientists are seriously thinking about using advanced versions to repair cellular damage and attack cancer cells. Also, robots are playing major roles in performing surgery.

Robotic aircraft less than five feet in length are now capable of plotting their own courses and flying routes thousands of miles long.

At the same time, these advances in robotics are blurring the boundaries between humans and robots. The same artificial "muscles" and "fingers" available for robots can also be used as replacements in sick and injured people.

The new edition of *Robotics* has been updated throughout, complete with new and larger numbers of photos and line illustrations. Compared to the original work, the chapters in this edition contain significantly expanded sections of material, including:

Chapter 1 artificial life has been added to the robotic family tree, and the introductory artificial life section has been expanded.

Chapter 2 a new section on animatronic robots.

Chapter 3 an expanded section on computer-managed production. The material includes computer-integrated manufacturing, just-in-time manufacturing, and flexible manufacturing. The hazardous environments section now includes the World Trade Center environment, new underground inspection robots, and new robots that perform live electric transmission line maintenance. The medical sciences section now covers robotic surgery and robots that work in laboratories that decode the human genome.

Chapter 4 new sections on fuzzy logic and robotics to deal with uncertainty, decentralized parallel processing, and grid computing.

Chapter 5 a new section on social robots that assist the handicapped, expanded sections on telepresence (haptics), and nanorobots.

Chapter 6 extensively rewritten, this chapter now covers Soviet Salyut space stations, U.S. *Skylab* space station, the Russian *Mir* station, and construction of the *International Space Station* (both U. S. and Russian sections), updated information on solar probes and observatories, and aerial robotics.

Chapter 7 new sections on humanoid robots; robotic toys, including Aibo; robot kits, including the Lego Mindstorms kits; and small-robot battles and contests.

Chapter 8 new sections on swarm (or ant colony) robotics and biomimetic robots, and an expanded section on artificial life.

There is also a set of appendixes that include an up-to-date list of recommended readings, website sources to check out, and an extensive glossary of terms used throughout the book. Readers may also refer to www.ellenthro.com (the author's website) for additional robotics information.

1

THE ROBOT FAMILY TREE

hat is a robot? An explorer of the ocean floor? A pioneer in outer space? An autoworker? A surgeon's assistant?

A robot can be all of these—and more. Of course, a robot is a machine, but it is also an independent being. For centuries the idea of the robot has included our powers to think, create, and control—the skills we believe make us unique among the creatures of the Earth.

The modern definition of a robot classifies it as a machine with these parts:

- a program that tells it what to do
- an arm, hand, or other moving part that performs a useful action, like lifting, assembling, or moving something
- one or more sensors so it can tell the difference between what it's working on and everything else
- a decision maker to decide whether to perform the action
- a controller, which is in charge of overall operations

Today's robots are at work in factories, hazardous environments, hospitals, laboratories, and in the home throughout the developed world. Research going on right now will produce robots with even greater skills and, increasingly, intelligence.

Robots through the Ages

The first robots on the robot family tree go back thousands of years, beginning with dancing dolls, or marionettes, and puppets in ancient Java, Japan, China, and Egypt, and talking statues in ancient Greece. The other three physical branches are much younger—factory machines, computers, and artificial life.

These nonrobotic, but real, ancestors are just part of the family tree. Robots have fictional ancestors too: androids, robots that are practically human, and cyborgs, beings that are part human and part robot.

So far androids and cyborgs live only on TV and movie screens and on book pages. They express the hopes and possibilities for "real"

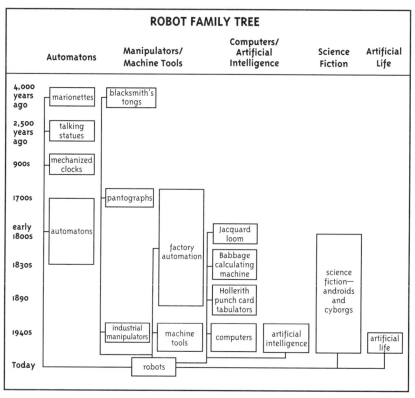

Robot family tree

robots. They also present the fears people have about robots. And they make all of us ask a basic question: What does it mean to be human?

DANCING DOLLS AND SPEAKING STATUES

Marionettes are puppets with strings attached to their joints, so that a person can make them move. They've been used as actors in plays put on in miniature theaters, with the people who pull the strings concealed behind the scenes. To an audience, the dolls behave like people. Strange as it may seem, marionettes are related to robots.

Marionettes designed to sing, clap, and dance have been found in Egyptian burial pyramids. Archaeologists tell us that some of these "dancing dolls" were used symbolically and realistically in religious burial ceremonies. In these, the deceased's body was sent on a voyage to the place of the dead. Some dolls represented the survivors, acting for them by sending food along on the death journey. Still others symbolized the transfer of money and power from the dead person to the survivors.

Puppets and marionettes were sometimes linked to various deities and were accepted literally by people of all ages. Even today, with a small stretch of the imagination, the puppets seem to come alive. The idea lingers, at least for small children, in stories like Pinocchio, in which a puppet becomes a living boy.

In ancient Greece, temple statues that "talked" conveyed oracular messages—lawlike prophecies or answers to questions from ordinary people as well as priests or priestesses. Perhaps the best known of these was the Oracle of Delphi. The oracle was variously meant to be the temple statue through which the utterances were spoken, the utterances themselves, or the deity to whom the temple was dedicated.

The oracular voices were usually those of a priest or priestess concealed below the statue. Sometimes they spoke in an overly emotional style. People believed that by talking, the statues represented life, or even that they really were alive. This may seem like a scam to some of us today, but the process extended the idea that a deity spoke directly through the clergy or even disguised itself by taking on the persona of the priest or priestess.

The oracle was first known in connection with worship of the earth-goddess. Speaking through the statue may have represented speaking through the Earth itself, especially in an earthquake-prone region like Greece, where the earth "spoke" powerfully.

In 2002, scientists discovered two intersecting fault lines directly below the oracle's chamber. Ethylene fumes that the scientists detected rising from the faults may have caused the oracles' over-emotional utterances.

Automatons

Robot is a 20th-century word and modern computer-driven robots are products of that century. But people have been building self-operating systems for many centuries. They are called automatons, powered by air, water, or mechanics.

True automatons, or self-operating systems, were first made and used about a thousand years ago in the Middle East and Europe. Mechanical clocks were popular throughout the Islamic world from the ninth to the 14th centuries. By the 15th century, Europe was filled with elaborate church clocks featuring people moving arms, lifting hammers, and striking bells, and even whole moving religious processions. The clock overlooking St. Mark's Square in Venice, Italy, is one famous example.

By the 16th century, smaller mechanical clocks ticked, chimed, and moved on walls and tables throughout the royal courts of Europe, the Near East, and Asia. But keeping track of time was just the beginning. There were mechanical drink dispensers, rolling down the banquet table and stopping at each place. And there were mechanical garden fountains. There was an automated washbasin that filled with water when someone came near, then extended a hand holding a piece of pumice stone (instead of soap). When the person picked up the stone, the automaton drew back its hand. And there were mechanical puppet theaters, complete with singing birds, hissing dragons, and moving people.

ALIVE AND NOT ALIVE

What is it about the human body that makes it alive, while a lawn mower or car isn't? The ancient Greeks thought it was the ability to speak. Today, scientists consider being alive to mean the ability to transfer genes from one generation to another. People inherit genes from their parents and pass them on to their children. The same is true for cats, trees, and even bacteria.

But there was a time when scientists pondered the idea that people might be alive because they were able to move around. That thought made sense in the 17th century, when modern science began. Galileo

Galilei, the Italian astronomer and physicist, codified the motions of objects on Earth and, with the aid of the first astronomical telescope, confirmed the Polish astronomer Nicolas Copernicus's conclusion, a century earlier, that the Earth moved around the Sun. The English physicist Isaac Newton, in his master work *Principia*, developed three laws that generalized all instances of motion, both where one body was at rest in relation to another (such as a person sitting in a moving cart) and where both bodies were in motion (a person walking on the moving world, for instance).

In fact, motion seemed to be so pervasive that scientists decided it was the guiding principle of the universe. It was just a short jump to the idea that motion was also what made people alive.

In the 17th century, interest in automatons increased and changed. The word automaton was coined then. People began asking, "Can we design and build machines to move so naturally that they are actually alive?" A scientist in those days who wanted to build what the 17th-century philosopher Thomas Hobbes called an "artificial mann" that moved might have asked several questions about it. Would it be alive? Could it think? Have emotions? Tell right from wrong? Be creative?

By the early 18th century, inventors were trying to answer these questions by building automatons that were amazingly lifelike. They were based on another branch of the new science—anatomy. If we make the automaton anatomically correct, they thought, there is a good chance that it will be "alive." Philosophers, alchemists, magicians, priests, and even kings all took part in automaton building.

Two of the most lifelike of these automatons were a flute player and a duck, both the creations of an 18th-century French inventor, Jacques de Vaucanson. His flute player was a man-sized figure of a faun—a mythical creature half-man, half-goat—seated on a rock and playing a flute. The most amazing thing wasn't the way the flute player looked, but what it did. It could hold any flute. It could take air into its lungs, blow into the flute with several different lip movements and tongue placements, and cover the flute's holes with its fingers to play a tune. A turning cylinder covered with raised bars, like those used in music boxes, controlled the whole thing. The bars in turn pushed levers, rods, and chains to produce the flute player's movements.

Vaucanson was experimenting with more than the body of his automaton. According to a common 18th-century belief, by giving his creation the ability to make music, he was also giving it a human voice—the voice of emotion.

This 1738 engraving shows three automatons made by Jacques de Vaucanson; the duck is in the center and the flute player on the right. [Engraving by Gravelot, from *Jacques Vaucanson, Mécanicien de Génie* by André Doyon and Lucien Liaigre]

Vaucanson also built an automated duck that quacked, drank, went into the water, lifted its feathers, spread its tail, and flapped its wings. It even had a ducklike digestive tract. The duck could take grain from Vaucanson's hand, swallow it, digest the grain (dissolved in water) in its stomach, and excrete the waste through an anus with a sphincter muscle. The inventor even left the digestive system partly exposed so people could see how it worked.

Vaucanson's creation greatly impressed the French king Louis XV. At one point, Vaucanson planned a working model of the human circulatory system. Louis offered to send him to South America, so he could experiment with a new discovery, called rubber, for the model's veins and arteries. Vaucanson apparently declined the offer, however, and never completed this project.

LATER AUTOMATONS

Many automatons that mimicked living systems were built during the 18th and early 19th centuries. Their main purpose was to entertain the public and, of course, to make money. Windup automatons played the organ, wrote letters with pen and ink, and drew figures with a pencil. One, called the Great Magician, even performed mind-reading tricks. A French father-and-son team named Jaquet-Droz built various automatons that played instruments, "wrote," and "drew," all controlled by music-box disks.

Sophisticated Parisians loved the lifelike machines. But not everybody was impressed. The historian Linda Strauss has found an 18th-century tale about a village woman who mistakes an automaton for a very talented person. When the woman discovers its true identity, she immediately decides it isn't so talented after all. Anything that an automaton can do isn't much of a human talent!

Probably the most famous—and mysterious—automaton was a chess-playing machine called The Turk, created in Vienna, Austria, in 1769 by Baron von Kempelen. Kempelen exhibited a chessboard-topped box, presided over by the life-sized figure of a man dressed in a turban and gown. Controlled by levers and pulleys, The Turk played chess against human opponents and usually won. Kempelen always began his demonstration by opening the cabinet, to show that no one was hidden inside. But questions were raised for decades. Was it really a smart automaton? Or was The Turk's expert-level game really controlled by a chess-playing child, midget, or amputee? Or by a series of them over the years?

Many people tried to explain the machine's success or expose it as a hoax. The question was never answered to everyone's satisfaction.

Interestingly, chess has always been considered a true test of computer programming skill. So The Turk was a forerunner of modern, "intelligent," computer chess-playing programs.

The Turk, a chess-playing automaton, was created in 1769 by Baron von Kempelen. (National Archives)

Automatons were amusements for the aristocracy. The 19th century was one of social upheaval and more practical priorities, and by the end of it, the powerful aristocracies and their amusing automatons had mostly disappeared. Even so, a few automatons were built as late as 1939.

At the same time, two other branches of the robotic family were growing and expanding—computers and factory machines.

Computers

At the same time that The Turk was mystifying people in the early 19th century, the modern computer had its beginnings. The location wasn't a scientist's laboratory, but a silk factory in Lyons, France. A weaver named Joseph-Marie Jacquard invented a punched-card system for weaving elaborate designs. The card holes controlled which warp threads were raised for the shuttle (weft) to pass under. This in turn determined the finished color and pattern of the fabric. Jacquard's system revolutionized the weaving industry. It is still used today around the world.

A few decades later, Jacquard's loom cards influenced Charles Babbage, an English mathematician. He invented a series of mechanical calculating machines. Babbage and his assistant, Ada, countess of Lovelace, laid the foundation for modern computer science.

Punched cards went electric in the late 1800s. The American Herman Hollerith invented a punched-card tabulating machine for use in the 1890 United States Census. Information from census interviews was punched into the cards. Then it was tabulated as the cards were run through the system. Each hole permitted an electrical signal to be produced and counted. Hollerith's company merged with the firm that eventually became the computer giant IBM. Updated punched-card systems are still used as, for example, voting machines, and the cards themselves are used as racetrack betting slips and advertising cards sent with Social Security checks.

Established companies and research institutions still have archives of data coded on punched cards and are very slowly converting them to computer files.

The first practical computers were built in the United States and Britain in the 1940s. They were huge building-sized machines that used bulky vacuum tubes to produce and change electrical signals. The transistor, invented in 1947 by the Americans John Bardeen, Walter Brattain, and William Shockley, showed that small pieces of silicon could do the same thing. Teams of transistors called integrated circuits—computer chips—led to today's powerful computers in tiny packages.

Factory Automation

Automation—the technique of making a process automatic—is a word that means different things to different generations. The word was coined in the late 1940s to describe machines doing the same factory work that people were doing. But the idea has been around since the Industrial Revolution (18th and 19th centuries). Huge water- or steam-powered factory machines replaced hand labor. The jacquard loom was an example. Yet those machines required large numbers of workers to tend them.

A century ago, the automobile automated the horse-drawn carriage. But the driver had to provide physical force to crank (start) the engine, shift gears, apply the brakes, and steer the vehicle. A half century ago, power steering, power brakes, and automatic transmission automated cars further.

Self-defrosting refrigerators automated the process of hand chipping the ice that built up inside. The old refrigerator, in turn, had automated the box that required hand loading of a block of ice every day.

There have also been several generations of factory automation. The assembly line, popularized by Henry Ford in the 1910s, was a big step forward. Building a car had been a complex process. It required skilled workers who could make decisions. The assembly line made factories more efficient by breaking the process into a series of simple tasks. The workers needed only limited skills and didn't make decisions, so they could be paid much lower wages.

This change led to charges that workers were being turned into "machines." Labor union demands and protective laws improved both working conditions and wages in many places.

At the same time, this system led to the next step in factory automation. Machines alone could do some of the simple tasks, replacing workers entirely. These machines came to be called machine tools.

Factory Tools That Do People's Work

In the United States today, anything made from two or more pieces of metal, wood, plastic, or other materials was probably produced with machine tools. These machines cut, drill, and grind. They turn out many identical copies of the same item. The hole is always drilled or the pieces welded in the same place, just as if a person carefully did the work by hand.

The most accurate machine tools employ numerical control—a system that uses numbers to describe the shape of the part and the tool's movements and working speed. Though there are even computer languages just for numerical control machine tools, they perform only simple tasks in comparison to the next generation, industrial robots, which perform more complex jobs.

INDUSTRIAL ROBOTS

Some people ask, are industrial robots just fancier machine tools? Or are there real differences? There are differences, even if they're not always clear. In fact, the same machine can be called a robot in Japan and a machine tool elsewhere. But in general, robots are more flexible, precise, and accurate, and they make decisions.

Industrial robots generally have guidance systems that put them at just the right place in front of their working materials. Their sensory systems tell them whether conditions are right to perform the task. Most robots are programmed to take the next step, deciding whether to go ahead with the task.

Factory automation today depends on two things—decision making and flexibility. A modern factory doesn't make just one product all the time. It must quickly change from one model to another, or even from one product to another. Robots let many manufacturers compete in the world marketplace.

Today's robots are still the first generation, the first products of all four main branches of the robot family tree. In factories they work as welders and painters. They assemble products and perform pick-and-place jobs (picking up materials and placing the finished product somewhere else).

Outside of the factory, robots must be flexible enough to move around on varied landscapes for firefighting and other hazardous or changing situations. Tomorrow's robots will be more independent, intelligent, and perhaps more human.

As in the past, people today ask, What's the difference between the movement of machines and the movement of people and other living things? Their answers still depend on three things: the technology of the day, their knowledge of how the human body works, and their beliefs (philosophy and religion).

Some of the most interesting answers come from computer scientists and fiction writers.

Artificial Life and Intelligent Robots

Some cutting-edge computer scientists have created what they think are living creatures in cyberspace. They are virtual life organisms that live, reproduce, and die only in the computer. The cybercreatures use computer programs instead of DNA to control their life cycle.

If these organisms really are alive, are they a new way to create androids or living robots? Are they the newest branch on the robotic family tree? Right now, only a few scientists say yes. Even artificial life researchers don't agree. A few say that reproduction constitutes life. Others argue that like living creatures in the physical world, artificial forms must also compete within the environment that formed them, proving their fitness to survive. (The topic will be discussed more fully in the final chapter.)

FICTIONAL ROBOTS

Authors of speculative fiction (also called science fiction) explore the following ideas of living robots in their TV shows, movies, and books:

- When does a robot behave so much like a person that it, too, is human?
- Do robots know what they are?
- What human qualities do they have? Can they acquire those qualities? Or get rid of them?
- Do almost-human robots threaten humans? Why? How do we deal with them?

The oldest robot-related story is probably the ancient Greek myth of Pygmalion, the king of the Mediterranean island of Cyprus, who sculpted the statue of a woman, which he called Galatea. It was so beautiful that he fell in love with it and prayed to Aphrodite (Venus in Latin), the goddess of love, to bring it to life. She granted his wish. This story intrigued people into the 20th century, and an updated version (with a store window mannequin instead of a statue) was even made into the Broadway musical and 1940s movie *One Touch of Venus.*

Two stories involve beings handmade of natural materials. In 16th-century European Jewish legend, the Golem was an automaton made of earth by a man who gave it life using a combination of faith and magic. In 1818, Mary Godwin Shelley wrote the novel *Frankenstein,* about a scientist of that name who created a creature out of body parts. Frankenstein then brought it to life. The creature was intelligent and sensitive. But it was also hideously ugly, so people rejected it. In revenge, it turned killer.

The Golem and Frankenstein's monster tell us about ourselves, as well as about robots—our desire to be physically stronger than we are, to show that we are as powerful as a god by creating life, that we think we can fully control these simpler, but more powerful beings. The more thoughtful of these stories, such as *Frankenstein*, also show that in creating such a creature, we inadvertently include human emotions and, ultimately, neither the creatures nor we are as powerful as we'd like to believe. For this reason, these stories live on into our time. Both have been made into movies and borrowed by modern authors.

The 19th-century American author Edgar Allan Poe is called the father of the modern mystery story. But he was also an early science-fiction writer. His story "The Man That Was Used Up," written in 1839, describes a human whose damaged arms and legs were replaced by mechanical substitutes. The character was a forerunner of the cyborg, and his mechanical limbs were forerunners of the "intelligent" artificial limbs now being developed for handicapped people.

A late-19th-century American writer of dark and edgy fiction, Ambrose Bierce, wrote a robot story called "Moxon's Master." It features an android that strangles its maker, who had defeated it in a chess game.

L. Frank Baum began writing the Oz series of books, starting with *The Wonderful Wizard of Oz*, more than a century ago. The books are filled with automatons, androids, and cyborgs. The most famous are the Tin Woodsman and the Scarecrow, who also appear in the classic movie *The Wizard of Oz*. The Tin Woodsman is really a cyborg that began as a human. He replaced his human body parts one by one with parts of tin, following a series of witch-induced accidents with his ax. The Scarecrow came to life as it was being constructed.

Other "live" Oz automatons with human qualities are Scraps, the patchwork doll; the Sawhorse; and the Gump, a flying couch. The Glass Cat had the human quality of compassion because it had a heart. Tik-tok, the mechanical man, was intelligent because it had brains (made of stainless steel). It spoke, thought, and acted logically, at least when it was wound up. But it wasn't alive. Tik-tok was proud of that fact even if this did let the "live" automatons feel superior.

The word *robot* appeared in 1921, coined by Czechoslovakian writer Karel Capek for his play *R.U.R.* (which stands for *Rossum's Universal Robots*). *Robot* is taken from the Czech word for forced labor. In the play, Rossum's Universal Robots look like people and have the human ability to kill. But they lack two things: a history and the imperfections that go along with being human. Eventually they kill off the entire human race. However, a chemical exists to give the robots emotions. At the end of the play, a male and a female robot

Human against robot—a still from Fritz Lang's film *Metropolis* [Photofest]

have become human themselves, the founders of a new, intelligent race.

Only a few years after *R.U.R.* the first movie about robots came along. It was *Metropolis*, a 1926 silent film by the German director Fritz Lang and today considered a film classic. The story tells of a future when factory workers have become little more than machines in the production system. The workers try to rebel, but a robot built by the evil factory owners leads them to their destruction.

EARLY ROBOTS

At the time that fiction books and movies about robots were gaining popularity, science was moving ahead. In 1927, engineers at Westinghouse Electric and Manufacturing Company began working on a series of mechanical men. The first, called Televox, was created that year. In 1932, Willie Vocalite was "born." He toured the United States, entertaining and astounding people just as the earlier automatons had done.

After Televox came Elektro, who appeared at the 1940 New York World's Fair with his dog, Sparko.

Both were the creation of Joseph Melton Barnett, a Westinghouse engineer. When Barnett spoke, Elektro obeyed. The spoken words caused vibrations that were converted into electrical waves, which then raised a shutter in front of a lamp and sent a flash of light to the photoelectric tube, or "electric eye," that served as Elektro's brain. This light was then converted back into an electric current, which started Elektro's motors. In this way, Elektro walked, spoke, and performed in other ways, making robots seem like a reality to those who saw him.

THE 1940S AND 1950S

During the 1940s and 1950s, three developments made modern robotics a reality: computers, artificial intelligence, and the study of systems.

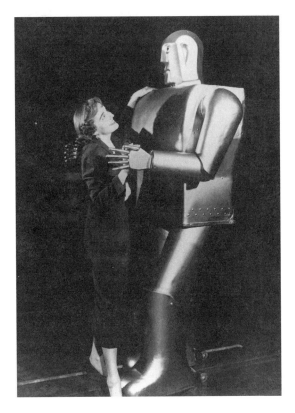

A futuristic dance partner? Elektro entertained visitors at the 1940 New York World's Fair. He could walk, move his head, arms, and fingers, count on his fingers, distinguish colors, and smoke. He could also speak 77 words. Weighing 280 pounds and standing seven feet tall, Elektro was the latest of several mechanical men Westinghouse built, starting in 1927.
[Westinghouse Archives]

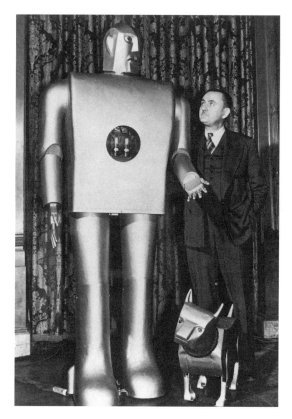

Elektro is shown here with Sparko, his "canine" companion, and their creator, Joseph Barnett. Sparko was a 65-pound Scottish terrier made of aluminum, steel, motors, gears, and switches. Standing a little over a foot tall and measuring 29 inches long, he ran behind Elektro, sat, stood up, wagged his tail, cocked his head, barked, whined, and growled. (Westinghouse Archives)

At about the same time that computer development began in the early 1940s, the British mathematician Alan Turing and other computer developers began to ask the question, can computers actually think for themselves? It was the beginning of artificial intelligence studies.

Also in the 1940s, the American mathematician Norbert Wiener and other scientists began studying systems. They studied how the human body's vision, motion, brain, and other systems resembled machine systems that did the same work. This was the field called *cybernetics.* It grew into several modern fields, including cognitive science (the study of human and computer thought). Robots—not automatons or "mechanical men"—controlled by brainlike computers and possessing some sort of intelligence now seemed possible. In actuality, it would take at least several decades before any sort of "intelligent" robot was developed. In the meantime, fiction hurried to catch up with theory.

Modern Fictional Robots

The science-fiction stories of Isaac Asimov in the 1940s and 1950s, such as *I, Robot*, explored the benefits of robots in a future society. Asimov wrote what he called the Three Laws of Robotics. According to the laws

1. robots are allowed to protect themselves,
2. but they mustn't disobey orders from humans, and
3. they must not harm humans.

Many computer scientists began using these fictional laws as they developed real robots.

Other writers have used this background to explore what it means to be human in a robotic world. For example, the hero of Anne McCaffrey's 1969 book *The Ship Who Sang* is Helva, a physically deformed baby who could never live on her own. Helva was given freedom, mobility, and the chance to express her intellect and emotions by being permanently grafted into a spaceship.

Movies also examined the effects of humans and robots on each other. Like people, some film robots were evil and some were good. For the first time, some robots were even lovable.

The first robot to capture the public's affection was Robbie, in the 1950s film *Forbidden Planet*, a science-fiction version of Shakespeare's play *The Tempest*. The good-natured Robbie was handy around the human outpost on an alien world, as well as being a protector and skilled worker. He was followed on the screen by the three plant-tending robots Huey, Dewey, and Louie (named for Donald Duck's nephews) in the 1972 *Silent Running*.

Probably the most famous of the lovable and intelligent robots were the *Star Wars* films' R2D2 and C3PO, from the late 1970s and early 1980s. The 1980s also gave us Number Five, an advanced sentry robot brought to life by an electrical accident in *Short Circuit*—a favorite of real-life robot scientists.

The Robbie-like robot on *Lost in Space* and the human cyborgs on *The Six Million Dollar Man* and *The Bionic Woman*, accompanied by Max, the bionic dog, brought the same admirable qualities to TV. (Hymie, the bumbling but earnest robot in the 1960s comedy series *Get Smart*, was played strictly for laughs—and satire.) The 1990s gave us Commander Data, the android Star Fleet officer on *Star Trek: The Next Generation*, who constantly tries to understand what it means to be human and to develop human qualities himself.

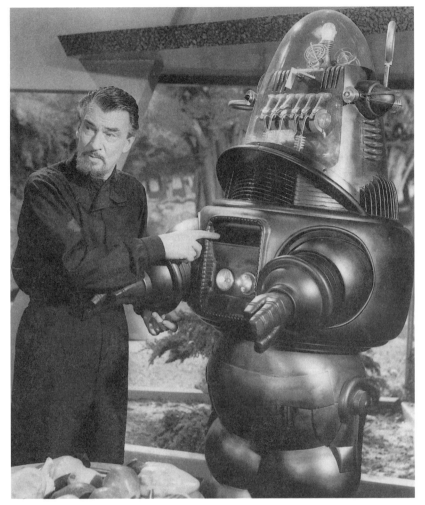

Robbie the Robot, with his human master, from the film *Forbidden Planet* [Photofest]

All this goodness has been balanced through the years by evil robots and cyborgs. *Star Trek: The Next Generation* also gave us the Borg, cyborgs who assembled themselves into machinelike beings to conquer the galaxy. Earlier TV series brought us the Darth Vader-like Cylons on *Battlestar Galactica* and the Daleks in the British *Dr. Who* series.

The lethal gunfighter robot of the 1973 film *Westworld* was a commentary on modern recreation and on movies about the Old West. Yul

Brynner, who had previously played gunfighters in several westerns, played the robot. Arnold Schwarzenegger's cyborg in the first *Terminator* movie was evil, but the cyborg in the second film was on the side of good. Both films were box-office hits.

A movie that explores the tension between androids and humans is *Blade Runner* (1982, from Philip K. Dick's novel *Do Androids Dream of Electronic Sheep?*). It is set in a future where almost-human androids, created for slave labor on a space colony, have revolted and killed people. Now some of them have come to Earth, where they're hunted by special police and killed.

Why did they revolt? One of them, who is winning a fight with a cop (Harrison Ford), says that now the human should understand what it means to be a slave, because he now understands ever-present fear. The androids' self-awareness, emotions, and memories make them seek the same freedom that humans have. But the humans have found a way to remain superior—these androids live for only four years.

The 2001 movie *A.I.: Artificial Intelligence* takes androids' emotions a step further. It asks whether the ability of a human to return an android's love lets the android come to life. Stanley Kubrick originally developed the film, from a 1969 story by Brian Aldiss, "Supertoys Last All Summer Long." After his death, Steven Spielberg completed the film.

Using elements from Carlo Collodi's children's story *Pinocchio*, *R.U.R.*, and even the jazz classic "Nature Boy" (by Moondog), *A.I.* shows us a future in which there are few humans. Androids, called Mechas, replace humans in many phases of life. Most couples are childless. For one such couple, a child android, named David, provides the ultimate human experience—to give and receive love.

From Robot to . . . Human?

The human desire to use the power of our knowledge and imagination to create lifelike beings has given us automatons and fictional robots. So far, we lack the ability to make real living robots. We cannot yet make beings with genes to be inherited, let alone with our intellect and decision-making talents. Robots also lack the mixture of good and bad qualities that religion and philosophy define as "human."

But scientists and engineers want the next generations of useful robots to include advanced computers and artificial intelligence—to be

Yesterday's robots are now collectibles of museum quality. These toys, on display at the Brooklyn (New York) Museum of Art, are from the collection of Robert Lesser. (a) A Meccano display robot, built in France around 1932 (b) A plastic-and-metal robot, made in the United States in the 1950s [Brooklyn Museum of Art]

smarter. They will make more decisions and they may seem more human. Already people are asking what the robots of the future will mean to our human skills, jobs, ethics, and freedoms. These are useful questions to keep in mind as you read the remainder of this book.

2

ANATOMY
OF A ROBOT

L ooking at one of today's working robots is not like looking into a
mirror. But robots and people are workalikes: To perform humans'
tasks the robots are designed with arms to lift and carry, hands as the
tool users, and legs (or wheels or treads) to move around. Human
senses and robotic sensors monitor the environment, and the brain or
a brainlike controller runs the whole operation.

Unlike people, most robots are designed to do only one or two specific jobs, so each robot has only the number of arms, hands, legs,
senses, and other abilities it needs for its work.

Arms, Hands, and Joints

One of the great differences between humans and other species is that
we can make and use lots of tools in lots of ways. A robot is the ultimate tool—one programmed to work independently. In fact, a robot *is*
a system of tools, its own most important tools being its arm and hand.
They are so important that they have their own branch on the robot
family tree, the branch entitled "manipulators/machine tools."

First, a word about the terms *arm* and *hand*. Some scientists and engineers don't like these *anthropomorphic* (humanlike) terms and use *manipulator* instead of arm and *end effector* instead of hand. This fits the idea of designing the robot for the task, materials, and environment.

Other scientists like arm and hand. They say that the human body is a good model for designing a robot and that once we have a good humanlike robot, later designs can become less "human." This was done in the past with airplane design, when birds were the inspiration. Early planes were much more birdlike than today's are.

The human arm and hand are made up of several body tissues. Bones for shape and structure and joints, tendons, and cartilage for extension, retraction, and rotation. The skin contains environmental sensors. The nerves bring instructions from the brain and send sensory signals to the brain. And the muscles use electrical and mechanical energy to do the work of motion, lifting, and carrying.

Robots' arms and hands imitate all these functions. The arm of a modern robot extends from a base (body) and is composed of rigid metal *links* that take the place of bone, tendons, and cartilage. It also has flexible *joints*—a "shoulder," a "wrist," and sometimes more. A computer, or *processor*, decides what to do; a motor or other *actuator* provides the muscle power. Sensors connect the robot and the environment. Electrical wiring carries messages between the computer and the arm and hand.

ARMS

Some experts say that the blacksmith tongs found in ancient Egyptian tombs were the first manipulators. People have always used hand tongs of some sort to handle hot or hazardous materials. The next generation was a tong that could be extended and retracted.

The need to handle radioactive materials in the 1940s brought a great advance in manipulators. These materials required very precise handling, such as cutting, pouring, and assembling. But they were too dangerous to touch directly or even get close to. The problem was solved with development of the *master-slave manipulator*. With this equipment, a person's (the master) arm and hand movements on a pair of handles were extended into a protective box containing the materials. The tool ends (the mechanical slaves) inside the box did the actual work.

At first, master-slave manipulators were powered entirely by human muscles. Then motors were added. They greatly increased the amount of work the slave end could do. Motors also allowed a greater distance between master and slave. In some cases, sensors at the slave end fed back information about operations to the master.

Computer control of the system was the next improvement. Commands were entered in a computer at the master end. They were sent to another computer that directed the actions of the slave. However, a person was still in control of the processing. The master could make changes in operations on the basis of personal observation.

In time, processing, too, was computerized, allowing many details to be handled automatically. For example, if a manipulator were placed in position to loosen a bolt, a computer program could direct the actual operation. This would make the control system a two-level operation, or *hierarchy*, with a person directing overall operations and decision making on the upper level and the computer handling the simpler tasks on the lower level.

Two kinds of joint are commonly used. One type, called *revolute* or rotary, is similar to the human kind. Others, called *prismatic*, work by sliding, in which a link is moved in space like a sliding door. But it keeps its direction relative to the link at the other end of the joint—a movement called *translation.*

For a robotic arm to move, the controller program must calculate several positions and routes. One is from the hand's present location to the point where it will do its job. This requires movement at each joint. You can see how this works with your own arm and hand. If you move your hand from this book to a can of soda, the positions of two joints—your elbow and shoulder—will change. For a robot to make this motion it must calculate the position change for each joint.

The flexibility of a robotic arm is determined by its *degrees of freedom*, or DOF. Each joint usually represents one degree of freedom. Commercial robots have from one to seven degrees of freedom. Experimental robots have as many as 20 DOF.

HANDS

The human hand is marvelously flexible. From the main palm, a joint attaches each of four fingers. Each finger has two joints of its own. Our jointed opposable thumb is attached very close to the wrist, so it can move independently. People can form their hands into many shapes, from making a tight fist to spreading it wide like a paddle, scoop, or net. And they can form various finger combinations for specific tasks, like picking up a feather, writing with a pencil, holding a basketball, adjusting a microscope, or taking a sample of cake frosting to lick.

A robot's end effector has a very limited and specialized shape and function. *Grippers* are robotic hands designed to grasp and hold. Some work like the human thumb and first finger. Others aren't humanlike at all—magnets and suction cups, for instance. Some robots have two or more grippers attached to the wrist.

Also, the gripper may be designed to best grasp a particular type of object. For example, some grippers grasp the object on the outside (external grippers). But if the part is ring shaped, the gripper may be designed to hold it from the inside (internal grippers).

In some industrial work, a robot must use a series of tools. So the arm and hand must be designed to pick up a tool, use it, release it, and pick up another. If only one tool is required, it may be attached directly to the wrist. This is often the case with spray paint nozzles, welding torches, cutters, drills, brushes, grinders, and liquid cement applications.

MUSCLE POWER OF JOINTS

An actuator at each joint powers a robot's arm. Electric motors are the most common type. *Pneumatic* (air pressure) and *hydraulic* (pressurized oil) systems are also used.

Pneumatically operated robots are modern versions of the 18th-century flute player. The robot's arms and legs are really cylinders and pistons (plungers)—like a hand-operated tire pump. Compressed air is pumped into a cylinder, moving the piston. This in turn provides the limb's motion. The airflow and the motion are computer controlled. Hydraulic systems work much the same way, but with a fluid instead of air.

Pneumatic and hydraulic types are reliable, but they can have problems. Their pumping systems have to be at a distance from the manipulator. Hydraulic systems can have the same problems that the family car does; lots of hoses, valves, and pumps increase the chance of oil leaks or seepage. Pneumatic systems are harder to use accurately, and air pressure can wear down the seals that keep air from escaping.

Moving Around

Robotic motion requires legs and feet, or substitutes. A mobile robot must also have a navigation system.

Most mobile robots don't walk on two humanlike legs. In fact, most don't have legs and feet at all. Two or four legs (or wheels) may seem "right" to us, because that's what mammals use, but they're not very stable. Three-legged stools are always stable, even when the legs aren't the same length. A four-legged table almost always wobbles a little. Many insects use six legs, often keeping three on the ground at the same time for stability. Many robots have sets of three, four, six, or more wheels.

Others move on treads or tracks, which are really a set of wheels running inside an endless pathway. Some robots have two tracks; others use four.

The *pantograph*—a series of adjustable parallelograms—was invented in the 18th century for copying graphics in various sizes. It was later adapted for grasping objects on high shelves and was widely used in grocery stores. In the 1970s, the 3-D pantograph—one that can move in three dimensions—was invented. It is now used for robot arms, fingers, and legs.

A huge three-ton research robot built by Ohio State University uses six hydraulically powered pantograph legs to walk and climb over steep obstacles. Balance sensors and vision help the controller to

An adaptive suspension vehicle, built on the principle of a pantograph [Professor Kenneth Waldron, Stanford University]

choose the robot's path. The 3-D pantographs make the legs extremely flexible. Force sensors allow adjustable pressure. The legs can crush a large piece of machinery or move a fragile box.

Whatever the legs' tasks are, the design problems are the same. The legs must support the weight of the robot's body and its load. And they must allow the most stable motion, especially over ramps, stairs, or uneven ground, or when carrying large or bulky loads.

A few human- or animal-style robotic leg systems have been designed, but most aren't in practical use. Exceptions are the four-legged systems that move animatronic, life-sized models of dinosaurs and other prehistoric animals for educational and entertainment displays.

An animatronic robot is much more limited in its abilities than an industrial robot, but very realistic in appearance.

Today's Dinosaurs— Robots or Not?

Dinosaurs are all over the place these days, even though they died out 65 million years ago. They may be walking the lawns of your local natural history museum or creating havoc in the latest *Jurassic Park* movie. They're perfect subjects for robotics, as a way of making them as mobile and realistic as possible. Some of them are *animatronic* robots. But others aren't robots at all.

ANIMATRONIC ROBOTS

Mainstream robots are designed to express human abilities or human ambitions, like making objects or "thinking." An animatronic robot has the same robotic "bones"—arms and joints, for instance. But its purpose is to show a particular person or creature. It is the animated version of a figure in a wax museum.

Each figure is designed from the outside in, based on what is known about the subject's skin, body or facial features, and how it moves and talks or makes sounds. For instance, one manufacturer, Kokoro Dinosaurs, begins creating a Tyrannosaurus rex dinosaur by talking with paleontologists—scientists who find and study dinosaur remains. Once the dinosaur's shape is known in detail, designers construct a steel frame. Then the robotic systems' pneumatic tubing and electrical wiring are placed inside.

Animatronic dinosaurs [Kokoro Dinosaurs]

A robotic movement, such as taking a step, begins when a compressor supplies air to cylinders in the leg. As the air pressure increases, it moves the pistons that force the leg outward. The actual dinosaur movements are designed into the control box. The dinosaur's vocalizations (based on the creature's anatomy and behavior) are also built in.

Next, a designer creates a realistic polyurethane body and molds it onto the frame. Finally, this is covered with a textured "skin." To make the skin look like the real thing, it is applied while the model is actually moving.

DINOSAURS IN MOVIES

Some dinosaurs in the *Jurassic Park* films are partly robotic. To control costs, many of them are produced with special effects using a combination of live-action puppets, large and small models, and computer graphics. Many film's scenes show just part of an animal, and often only the visible part is actually built. Scenes also may switch between full-scale and small-scale models. The dinosaur's motion can come from computer graphics or from people pushing a model by hand. And some

Dinosaur figure in the film *Jurassic Park*   [Photofest]

action scenes—such as a dinosaur crushing a car—may not actually show the animal at all, but only the people's responses to it. The dinosaurs in the 2002 television movie *Dinotopia* were rendered entirely by computer.

Robotic Navigation

A person might describe where he or she lives this way: "I live north of the shopping mall. It takes about 10 minutes to drive there in normal traffic." Possibly without realizing it, this person has just used a system of navigation called *dead reckoning*.

Navigation—getting from where you are to where you want to go—can be an exact science. Today airplanes and ships can rely on star atlases, satellite signals, weather indicators, directional compasses, and other equipment to sail from New York to Cape Town or to fly from Los Angeles to Tokyo. The same equipment can guide a spaceship from Cape Canaveral to the Moon or farther.

Dead reckoning—navigating by knowing only speed and direction—hasn't been state of the art since the 15th century. But even with today's tools, dead reckoning has its place—for very short trips, such as those a mobile robot might make from one room to another or into a hazardous environment. To use it accurately, you need to know three things: exactly how fast you're going, your direction, and how long you've been moving in that direction. If a person, or a robot, makes a mistake in any of them, he, she, or it will soon be lost.

You might think that moving from one place to another within a building would be an easy navigation problem. But any obstacle can throw the mover off course, like having to detour around a wastebasket or bumping into a chair that's out of place. As a practical matter, most mobile robots in daily use follow electronic signals embedded in the floor.

Navigation and other robotic actions involve interaction with the environment. People know about their environment because of their five senses. Robots' sensors serve the same purpose.

Sensing Its Surroundings

Sensors are robotic versions of the human senses. Robots can have sensors that imitate the functions of four of our senses: cameras for vision, acoustical arrays for hearing, smoke and gas detectors for smell, and force and *torque* (twisting) sensors for touch. Robots aren't big on eating so there are no taste sensors. But they can have thermometers, to keep track of environmental temperatures.

VISION

Vision systems are useful for mobile robots and for those that pick up, assemble, shape, or cut materials. Cameras play the role of the human iris, lens, and retina. They use visible light to take pictures in black-and-white or color. Of course, taking the picture isn't the same as knowing what it means. When a person looks at something, the brain gives the image its meaning. The retina passes the image—a set of light waves that bounce off objects—to rods and cones in the back of the eye. Interpretation of the image starts with them. They tell dark from light and one color from another. They pass the information on to a part of the brain called the *visual cortex* for final *image analysis.*

This includes recognizing patterns, which are shapes that occur over and over. Boundaries are important too. There are also building edges, lines that separate light and dark areas, and distinct objects within one's field of vision. Using three dimensions with our side-by-side eyes also helps us understand what we're seeing. Using only one eye makes patterns, boundaries, and objects harder to recognize.

The flat 2-D image from one eye is fine for picking up a pencil from a table. But telling how close someone is standing to a grove of trees requires depth, or 3-D, vision.

But image analysis also has many uncertainties. Is that tall, dark object a tree or a telephone pole? Is that line a boundary between sidewalk and grass? Or is it the edge of a high wall? Are those rippling patterns a pond surface or a wheat field? Is that short, bulky object in the corner a trash bin or an air-conditioning unit?

To answer these questions, your brain must compare its first analysis with what it already knows. Our memory holds past experience, plus what we're learning now from books, pictures, and sounds. The brain uses these memories to decide what you're seeing and what, if anything, you should do in response.

Two people looking at a scene may not agree on what they see. One person may think the bulky object is an air-conditioning unit, while the other thinks it's a trash bin. Each person made a decision based on incomplete or uncertain information. After discussion, one of the observers may have a change of mind about the scene, or else both observers can decide more information is needed.

For a robot's vision system, the camera supplies the image. This is often a *CCD*, or *charge-coupled device*, which, like a digital camera, uses a light-sensitive computer chip. The brain power and image analysis come from a computer. The robot's vision analysis program is much simpler than that of a person's. Most robots can't deal with uncertainties the way that a person's eye and brain can.

A robot on the job must make yes-no decisions. Is the assembly piece in the right place to have a hole drilled in it? Is it close enough to pick up? Most industrial robots that "see" have two-dimensional, black-and-white vision systems. And they can't move over to get a better look.

Mobile robots usually have some three-dimensional vision. And some of them can see in color. They need 3-D and color because their vision is part of their navigation system—just as a person's is. Not only are they moving, but some of the people or objects they see may be moving too. Other portions of the electromagnetic spectrum besides visible light can also be used to take a picture. Some robotic eyes can "see" infrared or heat waves. Infrared waves come from living people and animals and from objects that are warmer than their surroundings. This is useful at night or where the light level is very low. Some robots are equipped with X-ray systems.

TOUCH, FORCE, AND TORQUE SENSORS

One of the most advanced touch-sensing systems ever developed is the fingertips and palm of the human hand. This system contains four kinds of organ to sense shapes and textures. It has 17,000 sensors. They work by detecting changes in the skin, such as stretching and pressure. Then they trigger nerve cells, which pass the information to the brain's *somatosensory cortex* for processing. Touching this book, for example, will activate them.

The "fingerprint" creases and ridges—the fingertip pads—contain two types of organ. When sliding a hand down a side of the book, one type of organ senses its edges, size, and shape. Rubbing the finger pads over the book edge activates the other type of organ, which is also found in the palm. It provides information about the page's texture. Pressing a hand down on the book activates a third type of organ, telling how rigid or flexible the surface is. Finally, stretching the fingers out flat makes the skin vibrate and activates the fourth type of touch organ.

Robots aren't always accurate. The design, the materials they're made of, and heat and other environmental conditions can cause navigational errors. Robots need touch sensors to tell the controlling computer whether the robot hand is actually where it should be. The controller must calculate the difference between where the robot actually is and where it's supposed to be. Then it sends a command to correct the error.

Scientists haven't yet been able to develop a material for robotics that's as flexible and sensitive as human skin. But they have developed sensors that can tell how an object's shape and contour respond to compression (force) and distortion (torque, or twisting motion).

Position can be sensed on the actuator by one of several devices. The sensors measure changes in electronic signals caused by rotation of the motor's shaft. The most common types are the *incremental rotary optical encoder* and the *resolver.*

The human hand can control force very precisely. For instance, it allows just the right amount to grip an egg, which is different from the amount needed to hold a tennis ball. Still another amount of force may be required to hold the aluminum cap of a soda pop bottle and twist it off. Too strong a grip will bend the cap out of shape. But a much stronger grip may be needed to twist a steel nut from a bolt.

Assembling materials, drilling holes, and lifting objects are common robotic jobs. The robot must use the right amounts of force and torque for the material. Those amounts are designed into the robot and its computer program for each job.

Force and torque are usually measured at the hand end with *strain gauges*. These contain semiconductors that produce signals that get stronger as more force or torque is used.

SENSING SOUND, ODOR, AND HEAT

When a bat flies through the night air, it emits high-pitched sounds that echo back after bouncing off trees, buildings, and other objects. Whales and porpoises use similar systems under water. Sound like this is called sonar (sound navigation and ranging). Sonar has long been used for ship navigation and submarine detection. It is the basis for some self-focusing cameras. And mobile robots also use it.

Sound is a disruption of the air or water that is expressed as waves of energy. Like light, sound waves can be reflected by objects they strike. The distance between a sound source and its target can be measured by the amount of time required for the echo to return. A robot can use sonar to tell how far it is from a wall or doorway. Robots may also be equipped with smoke and gas detectors or thermometers.

Command Center—The Brain

The human brain is a real jack-of-all-trades. It's the captain of a person's "body team," controlling all the operations. It can take totally new information, analyze it, and find memories (stored information) for comparison. It combines the information in new ways. And it learns from the experience. If it has to, the brain can act immediately on its new knowledge and then store the whole thing for future use. In computer terms, it maintains a data, or knowledge, base and performs information processing.

Your brain guides your interactions with your world in several ways. It learns, it stores knowledge, and by staying in constant touch with the world, it can plan your actions and tell you how to react second by second. A robot has to know its world too, in order to interact with it.

A Robot's World

Each person has a personal world that is part of the larger world. A student's world is made up of familiar areas—home, neighborhood, bus route, school, park—and the student interacts just with those parts that the senses can detect and the information that the brain can process.

Each species senses and understands its personal world differently. A human sees in a wide range of colors. A dog or cat sees mostly in black and white. A person immediately recognizes a blowing leaf by its shape. But a cat may see it as a threat. At night, a human sees darkness, but a cat sees distinct objects. The air that sounds quiet to a human may be full of sounds for a dog.

A robot moves in a world too. Its world is limited by its design, the materials it's made of, its sensors, and its programming. A robot's controller program has to store a picture or word-picture of that world. It must also know how the robot should interact with the world. It must keep up-to-date information about the way that it's actually interacting. And it must be able to send instructions to correct any errors.

Putting It Together in the Real World

Suppose a softball player is at bat in a game. Several experts in that person's "body team" are taking part. The eyes are at work watching the ball. They send a stream of information to the brain about the ball's speed, height, and perhaps the way it is turning. At the same time, the brain is evaluating this information. It is also controlling the way the person stands at the plate and calculating how the ball should be hit.

As the ball nears the plate, the brain decides whether the batter should swing or let the ball go by. If the brain selects swinging, it sends "how to move" messages to the arms, upper body, and legs. If it made the right decision, the batter hits the ball. In that case, the brain sends further messages, instructing the hands to drop the bat and the legs to streak to first base.

Sometimes the brain miscalculates, and the batter swings but misses. In such a case, the brain adds this experience to its memory to help make a better decision next time.

The brain has been working on several problems at the same time—what is called *parallel processing*. The human brain is very good at this. For instance, at the same time the batter's brain was deciding about the pitched ball, it was also making sure the person's heart was beating, telling the intestines to continue the digestion process, and directing every other body function.

Some of these tasks are so routine that the person doesn't realize they're happening. Other tasks require advanced processing. Making decisions and controlling operations are very complicated chores. There are lots of facts and lots of uncertainties. So the brain also works hierarchically.

Most computers can do only one job at a time, what is called serial processing. They're very good at processing numbers. In fact, they can do that much faster than the human brain can. But a robot may have several tasks to do at the same time. In this case, it needs a different serial computer for each task.

The computer controlling a robot's overall operations and making the decisions is a lot like a personal computer. It sits on a desk and has a screen and a keyboard. Other controllers are single purpose ("dedicated") microprocessors (computer chips). There is a different chip for motion and guidance, sensor data processing, sound, speech, and other functions. They all work together, so the effect is brainlike.

Robot Academy—Robots Learning to Be Robots

Factory robots must be taught to do the useful work, such as assembling a product or carrying an object from one place to another. The teacher first has to know the robot's abilities; how it learns and how much it can be taught. For instance, the teacher must know what the robot's arm can do. Industrial robots have either *fixed-stop* or *servo-controlled* arms.

FIXED-STOP ROBOTS

A fixed-stop is like a train that stops only at its stations, never in between. The arms of fixed-stop robots can stop only at preset points on the *work path*. This is the route the arm or each link takes from its starting point until it reaches the work location.

If a robot has fixed stops, it can learn only the sequence of moves. Robots of this type are used for simple *pick-and-place* operations. A person can teach one to pick up a finished item from a worktable, turn, and then place the item on a conveyor belt.

SERVO-CONTROLLED ROBOTS

A servo-controlled robot has more abilities. A *servomechanism* is a device that knows where a joint or arm is, in relation to its possible range. It feeds this information back to the control program. This lets the controller compare where the arm should be and how fast it's moving with its real position and speed. Once it knows this, the controller can stop the arm anywhere along its work path.

There are also two methods for managing servo robots' movements.

- *Point-to-point control* means that the robot can stop only at points it has already been taught. This type of robot is used for assembly tasks. For instance, it can go to a storage area, retrieve a part or tool, return to its workplace, and perform a task. It can go through this sequence of motions many times. Any errors that creep in can be adjusted by the controller.
- *Continuous path control*, where only the beginning and end points of a motion are set. Such robots perform work with a moving arm, such as painting a car body or welding two pieces of metal. The arm can "learn" by being guided through its motion, or the controller can calculate the actual path and any in-between stopping points.

Most industrial robots take only straight paths. More advanced robots can also take curved paths.

TEACHING METHODS

Someone must teach a robot what movements to make and when to make them. The process is much like teaching someone to dance or roller skate.

There are two ways to teach a robot. A human teacher can move the arm through the motion, recording each start and stop position with a control box called a *teach pendant*. This is called *lead-through* programming. Once all the movements are recorded, the robot can repeat them at work. This kind of robot is called a *playback* or *record-and-playback* robot.

Other robots can be programmed using a computer language specially designed to describe motions and angles. Most "simple" human motions are quite complex and difficult to describe. For instance, to pour a glass of juice from a pitcher requires grasping the handle of the pitcher, lifting it, and tilting it. The arm must move and hold the pitcher at a precise angle. When the glass is full, the pitcher must be tilted and set back down, again by moving the arm and changing angles.

To program a robot to do the same thing requires a language that tells the robot when and how far to move, grasp and release, and change the angle of one or more joints. Some of the languages also take sensor readings, so the robot will stop, for instance, if pressure on an object is too great. (In fact, pouring a glass of juice is beyond the abilities of today's robots.)

The latest kind of teaching is *graphics-based programming*. It combines the record-and-playback and programming methods. A computer simulation of the robot in action is created. The simulation is programmed with the work sequence. Motion and programming errors are corrected on the simulated robot. Then the final version of the program is transferred to the real robot's controller.

From Master-Slave to Android

Some of today's robots have very little independence. Others have a lot. Some are remotely controlled by a human operator, which is called a teleoperated device. They are direct descendants of the basic master-slave manipulators. Teleoperation provides very precise and realistic operations, almost as if the operator were performing the operations directly:

- An **intelligent teleoperated** system has some computerized control at the slave or remote end.
- With **supervised autonomy,** the robot's computer performs almost all the decision making as well as calculations, but a human can still take control of either operations or decision making.
- The **fully autonomous** robot is completely free of human control.

Some people might think that the more independent robots perform the most complex or sophisticated tasks. Despite the impression given by science-fiction androids, this is not necessarily true for today's robots. For instance, some of today's most sophisticated tasks, such as undersea exploration, are performed with intelligent teleoperated systems, which are also called *telerobots*.

We haven't progressed to fully autonomous robots yet, let alone androids. But in the next chapters you will see the other robot types at work (and play) today.

3

ROBOTS ON THE JOB TODAY

Right now robots are part of our lives, perhaps more than we think. TV sets and computers are often made by robots. The office mail in some large companies is delivered by robots. Nuclear power produces electricity in many areas, and robots perform maintenance work on the reactors. The family car undoubtedly bears the robotic touch: parts welded together, the windshield installed, the body painted. People can even buy a security robot to patrol the house or office building when nobody's there.

There are about 90,000 robots at work in North America and many more around the world. And the number is growing. When robots first came on the scene, people thought each robot would be a general worker. They thought a robot would be able to use many tools to do lots of tasks. Things didn't work out that way. Each type of job was too specialized. And the same job turned out to be very different in different plants. Experience has shown that robots must specialize. Each model must be designed for a single purpose. Even then, it often must be adapted to the environment it will be working in.

Some robots are designed for industrial manufacturing. They are very different from robots used in hazardous environments, delivery

Arc-welding robot [Courtesy of Fanuc Robotics, North America, Inc.]

and other services, and security. Most robots today are found in factories.

Robots in the Factory

In the United States, industrial robots are used chiefly in the auto industry. By themselves or teamed with machine tools, robots also work in general manufacturing. They drill and cut. They assemble and spray paint. They load and unload materials. They inspect.

In many cases, robots are used in traditional factories. But they have had only limited success. To be useful, industrial robots must be part of an overall automated and flexible system, controlled by computers.

Computer-Managed Production

Computers and robots are standard features in industry. Many plants worldwide use computer integrated manufacturing (CIM). This is sometimes called a flexible manufacturing system. During the 1990s, computers made it possible to use a *just-in-time* supply system. It saves time and money by ordering parts and supplies as they're needed.

COMPUTER-INTEGRATED MANUFACTURING

Actual product making is just a part of factory automation, even with robots at work. First, the finished product has to be designed. The manufacturing process must be planned, so it is accurate and efficient.

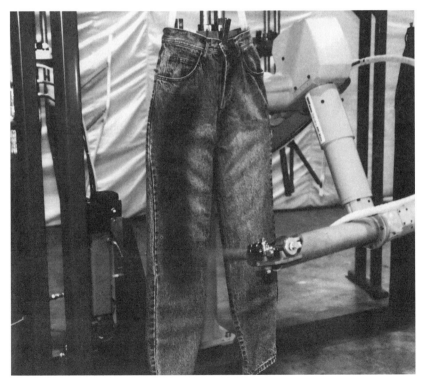

A robot makes new blue jeans look fashionably old with a local abrasion process.
(Courtesy of Fanuc Robotics, North America, Inc.)

Production and sales must be coordinated, so there aren't too many items to sell or too few for the customers. Finally, plant managers and owners need an accounting system. This keeps track of the money spent and, hopefully, the profits.

A single computer network links the ordering department, the design department, and the sales and accounting departments. Some companies also tie the computer system into the manufacturing and assembly process, including robots. There might even be automated carts and conveyor belts to move parts and materials between work areas.

JUST-IN-TIME MANUFACTURING

Computers also allow ordering of raw materials and parts as they track the manufacturing process. Once, these materials were ordered far in advance, then stored until they were needed. This meant that the manufacturer spent money on parts and supplies that might not be used for weeks or months—long before the final products were sold. "Just-in-time" ordering means the company buys them just before they're needed, often only the day before. This way, money is spent much closer to the sale of the finished product. Also, the company doesn't have to provide warehouse space for the materials, saving more money.

Suppose a manufacturer has brought out one of its products in a new color. The sales department learns from the retail stores that the new color isn't selling very well, but people are buying so many of another product that all the store shelves are empty. Once, the company would simply have to order more supplies and wait several weeks or longer for them to arrive.

With just-in-time ordering, the new supplies can be at the factory in a single day. At the same time, the manufacturing robots can be reprogrammed so the factory line can quickly shift to making the popular item. It is also fairly easy to gear up for a new and improved item.

FLEXIBLE MANUFACTURING

Robotic manufacturing systems are used for a wide range of products, from cars to contact lenses. The auto industry used to need almost a year to retool an assembly plant for new models. A large plant may have a thousand or more record-and-playback robots. Each one had to be reprogrammed by hand. The changeover itself was costly, because each model had its own specialized tools. And an idle assembly line meant products weren't being made and sold.

Flexible manufacturing systems have greatly speeded up the process and reduced costs. Nissan Motor Manufacturing Corporation U.S.A. was a pioneer in installing flexible manufacturing systems in its plants. The heart of Nissan's flexible system is a team of robots that can grip and weld any body parts from any of the company's car models.

After the work is done, still other sensor-equipped robots inspect it for accuracy. Information about manufacturing errors is fed back to the computers. Programming is then changed automatically, so work on the next piece is more accurate.

To switch models, only the robots' programming has to be changed. The company found that this cuts the model changeover time to just a few months, and the costs are reduced by 80 percent.

A flexible manufacturing system is installed at Nissan. [Courtesy of Nissan Motor Manufacturing Corporation U.S.A.]

In old-fashioned car making, design of the production process didn't start until the product's design was complete. With a flexible system, the product and the manufacturing process can be designed at the same time. When the product is ready to be made, the automated assembly line is also ready to go to work.

Service Industry

The service industries—those that provide a service to consumers, rather than a product—are using robots in many places, sometimes where you wouldn't expect them. Robots are becoming part of the service work team in restaurants and food service operations, equipment maintenance and floor care, and mail and delivery. For example, the U.S. Postal Service has added floor-cleaning robots to six large mail-handling facilities. The robots are cost-effective for the large floor areas involved.

In fact, robots are generally most useful in large operations, where they can do routine work, while people take on more specialized jobs. Mail carts and material delivery systems are widely used examples. Delivery carts are also used in automated warehouses, where most of the work is done by conveyor belt systems, elevators, and almost-robotic carts. At least one large hospital uses a fully robotic cart system for its deliveries.

ROBOTIC DELIVERY SYSTEMS

Most people think of hospitals as places where sick people get well and where women have babies. This, of course, is part of their purpose. But in its operation, a hospital is a big delivery system. Medications and supplies must be delivered from storage to nursing stations and operating rooms. Clean sheets and towels must appear regularly in patients' rooms. Soiled instruments and other leftovers must be taken to the laundry or the trash. Food—regular meals and special diets—must be delivered on time from the central kitchen to the patients. Afterward the trays of dirty dishes must be removed.

In many hospitals, it's common to see heavy trays of food or linen baskets being hand pushed onto elevators and then into public corridors. It used to be that way at the U.S. Navy's Balboa Hospital in San Diego. But in the mid-1980s, a new and larger hospital was built, and robots were designed into it. Instead of human power, robotic carts deliver supplies and food. They pick up soiled laundry and used trays. But you never see them in the elevators and corridors. This is because the hospital has separate corridors and elevators just for the robots. Following electronic pathways placed beneath the flooring, robots move through the hospital.

The three-wheeled robots look much like golf carts. They are powered by four 12-volt batteries and can carry as much as 500 pounds.

The heart of the system is the computer control room, complete with map that keeps track of all the carts. Dispatchers use keypads at "robot stops" to summon a cart to a pickup point. The cart positions itself under a basket of supplies or a tray of food, raises its back platform to lift the rack off the floor, and goes to its destination. Once the cart gets its instructions, it moves slowly—about a half mile per hour—through a robot corridor. It takes a robots-only elevator to the correct floor.

Robotic cart in a robots-only corridor at the U.S. Balboa Naval Hospital, San Diego
[JO1 Sonya Ansarov, USN]

Radio-frequency signals transmitted through guide paths buried in the floor provide all the commands to receptors located underneath the carts. Clusters of magnets buried in the floor generate magnetic fields that open or close switches on the cart, allowing it to send or receive certain types of signals. These include "go to the elevator," "switch to a different radio frequency," "turn," "go in another direction," or "unload." Programming and signals from touch and optical distance sensors keep robots from moving too close together. A cart will stop if it bumps into something. The optical sensors detect differences in light levels. The carts are highly reflective, and if a sensor picks up a high light level, it will issue a stop signal.

Each floor in the hospital has special rooms that connect to both the robot and the human corridors. Only certain employees are admitted. Otherwise, no one sees either the robots or their travel lanes. When a cart stops at a "robot stop" outside the room, an authorized employee unloads the cart and sends it on its way. Only then are the supplies and food distributed through the human corridors. Trash, empty food trays, and other throwaways follow the same route, only in reverse.

SAM—Sewer Access Module

Fiber-optic cable allows high-speed computer communications throughout the United States. But laying the cable to every home and office building is slow and expensive. It can also be inconvenient because trenches must be dug in city streets to lay the cable, then paved over.

The robotic Sewer Access Module (SAM) has changed things by laying the cable through existing sewer pipes. SAM looks like a snake on wheels. The remotely operated robot is six inches wide and a yard long and is made from a series of single-purpose units, or *modules*. SAM is inserted by hand through a manhole. From then on, a remote operator, sitting in a truck overhead, commands the robot's work as it makes a series of trips through each segment of sewer pipe.

First, its front-mapping module takes pictures of the pipe, so its cable-laying trips can be plotted exactly. Next, a ring-installation module replaces the mapping module. By remote commands, SAM slides a series of steel support rings snugly into the pipe at set intervals. Finally, a new module unrolls a narrow steel tube or conduit through the sewer pipe, clamping it to the top of the support rings.

The actual fiber-optic cable is then fed through the conduit by a nonrobotic pressurized air system. The final step, attaching the system to the home or office building, is done by hand.

SAM was developed by a Swiss firm, Ka-te, and first used to lay fiber-optic cable in Hamburg, Germany. The company's U.S. partner, CityNet Telecommunications, Inc., has laid cable in Albuquerque, New Mexico, and plans to use the system in other cities. CityNet says that SAM works more than 50 percent faster than a conventional installation and costs no more than digging up a street and repaving it.

Working in Hazardous Environments

In ordinary times, robots work in environments too hazardous for humans, such as nuclear power plants, bomb disposal, capture and restraint of suicide bomber suspects, and toxic chemical spills. After September 11, 2001, they took on new roles within the World Trade Center site.

THE WORLD TRADE CENTER ENVIRONMENT

After the September 11, 2001, attacks, the World Trade Center site became a hazard too dangerous for humans to enter, because of the thick dust and smoke, or because many areas were too small. Portable commercial robots, experimental prototypes, and quickly modified versions were rushed from many parts of the United States to sift the materials. This month-long effort, in association with fire department workers and rescue-dog handlers, was the first robot-assisted urban search and rescue effort in the world.

The robots themselves were developed only after the terrorist destruction of the Alfred P. Murrah Federal Building in Oklahoma City in 1995. A former U.S. Army officer and a rescue worker at the Oklahoma site, John Blitch, regretted the lack of any robotic support there. He coordinated development of small portable robots by various commercial firms and government agencies.

The Foster-Miller Corporation's Talon portable robot, for instance, is a modular system that travels on treads at four mph and can be controlled precisely for up to one mile by two-way radio frequency (RF) or fiber optics and as many as seven cameras with audio and data feedback. It can work in darkness as well as in the light. It has a 64-inch-long, two-stage arm with a gripper and can carry more than 200 pounds. At the World Trade Center it allowed rescue workers to evaluate remote areas and determine whether they should be searched further.

Inuktun Services Ltd.'s Micro Tracs are individually powered tread systems, about seven inches long and two and one-fourth inches high, ordinarily used for video inspections of piping and other purposes. Able to climb vertically inside pipes that are eight to 12 inches in diameter, one unit located several untouched interior rooms deep within the rubble. Further investigation by a firefighter revealed several bodies that were then recovered.

Other robots included Urbots (Urban Robots), provided by the U.S. Navy Space and Naval Warfare Systems Command, which were designed for the U.S. Army's 10th Mountain Division, in Ft. Drum, New York, to clear passages in potentially hostile urban environments, such as tunnels and sewers. Each remotely controlled robot weighs less than 40 pounds, can operate upside down, and can move both forward and backward. Urbots' video systems were used at the World Trade Center to investigate narrow spaces and show where they should be cleared.

PackBot, a prototype built by iRobot Corp. and funded by the Department of Defense, travels on treads that can also act like "flippers." These let the robot climb stairs and hills and stand upright for navigating narrow or twisting spaces, such as those in the World Trade Center debris. It can operate either autonomously or by remote control. According to iRobot, it can survive a three foot drop onto concrete and launches from a second-story window.

OTHER ENVIRONMENTS

Robots work in the electric-generating industry. They also help police get rid of bombs. When there's a toxic chemical spill, robots take part in the cleanup. They fight fire. Some have been designed for the construction industry. How do you inspect and weld steam pipes without turning the system off? Design a special robot that can successfully work in temperatures of several hundred degrees Fahrenheit.

Robots are designed for specific tasks so there is great variation in how they look. Some are huge. One looks like a giant earth-moving machine. Others are tiny. One looks like a snake, another like a six-wheeled roller skate.

WISOR—WELDING AND INSPECTION STEAM OPERATIONS ROBOT

In the late 19th century, many cities built central plants that heated office buildings with steam, carried in miles of underground pipes. Some plants are still working, but the pipes are old and worn. When they break, they cause floods and need expensive repairs.

In New York City, the utility company Consolidated Edison Co. of New York now uses a robo-welder named WISOR to inspect and repair the pipes. The remotely operated, eight-foot-long machine, created by Honeybee Robotics, Inc., can tunnel 135 feet into a pipe, where the temperature is 275 degrees Fahrenheit. WISOR has three major parts: a *milling machine* that cuts and shapes a broken pipe so it can be welded back together, a flexible middle section that actually moves the 700-pound robot through the pipes' straight and curved sections, and a rear welding section. WISOR's system also has four cameras that let the remote operator watch its progress on a video screen.

NUCLEAR REACTOR INSPECTIONS

The inside of a nuclear power reactor is filled with pipes and tanks that carry radioactive hot water or steam. A combination of radiation and aging can damage reactor parts. This could cause a leakage of radioactive materials into the environment. Reactor operators must inspect the reactor's insides constantly. Since the radioactive environment is too dangerous for people to work in, robots are used instead. Some robots patrol the interior of the reactor, inspecting pipe leaks. Some robots are small enough to roll or climb inside pipes that are only six inches in diameter. Robots can carry out inspections using video or sonar. Or they take gamma ray pictures (like medical X rays). A few have sound sensors that "listen" to the pipes. Strange noises can mean that a pipe should be replaced or bypassed.

Others retrieve tools or additional materials left behind during construction or repairs. Some are equipped with environmental sensors. They measure the air temperature, humidity, and radiation level.

Robots make repairs inside the reactors. And they clean walls, change filters, clean up contaminated pipes and other areas, and remove sludge. Some can even turn valves on and off.

LIVE ELECTRIC TRANSMISSION LINE MAINTENANCE ROBOTS

The lines that transmit electricity can be a very dangerous working environment. Workers must be protected against the electricity. They work high off the ground on towers or trucks with lifts. They cannot work during high winds or lightning or when it is very hot or very cold. Even when conditions are right, the workers must carry heavy equipment to the lines with them. They perform delicate repairs and maintenance, such as changing insulators. And they do it while wearing heavy gloves and other clothing to protect them from the electric current. Robots that work on live electrical lines have been designed and built in Japan and Canada, as well as in the United States and Spain. For example:

- **The Japanese robot.** The Yaskawa Electric Corp. and Kyushu Electric Power Co., Inc., have jointly developed a robot with multiple-axis arms that works on live 22 kV distribution lines. Yaskawa developed the insulated and weatherproof materials for it. Kyushu Electric now uses 70 of the robots for regular maintenance work. The company credits them for contributing to its record of zero blackouts and zero human incidents.

TOMCAT robot, designed and built in the United States in the 1980s, performing maintenance work on a live electric transmission line (Alliant Energy—Doug Wollin)

- **The Canadian robot.** Hydro-Quebec, the provincially owned utility, is currently testing its own design of a noncomputerized master-slave manipulator with five DOF for line-line maintenance. Once it's in regular use, the worker will stand below the line and control the slave robot with the six-DOF joystick on its master controller. The controller, called Kali, was developed by McGill University and can be upgraded simply by replacing the computer board. The robot is expected to double the efficiency of its maintenance workers.

Hydro-Quebec is also developing a second generation of robots, this one computerized, to further automate maintenance tasks.

Security Robots

Almost all office buildings, stores, and factories have security systems. These are human guards, video camera systems, wall-mounted alarms, or a combination of systems. Good security systems are expensive. Guards can't be everywhere at once. Video cameras and wall alarms aren't always successful in detecting intruders or other problems.

A sentry robot to patrol the building can be an added security tool. In fact, a leading manufacturer of sentry robots and automated security systems, Cybermotion, Inc., says that security managers report high turnover and low quality performance by security workers has been the main motivation for switching to a robotic sentry. A sentry robot is a special type of mobile robot. Besides navigating through its environment, it must have sensors and programming to tell normal from unusual conditions.

When a robot patrols the corridors, its sensors tell it that things are normal, such as office furniture being where it should be, corridor lights turned on but office lights turned off, and the building's temperature at 68 degrees. The robot notes a few ordinary abnormalities, such as the wastebaskets in a different place.

The robot is also programmed to observe out-of-the-ordinary abnormalities, such as a corridor temperature much higher than normal or smoke in the air. These conditions can cause the robot to sound a fire alarm. If the temperature is too low, it could mean an open outer door. The robot can tell if lights are on where they shouldn't be. Under these conditions, the robot may give an intruder alarm.

A CyberGuard SR3 security robot performing an asset inventory while patrolling an office complex [Courtesy of Cybermotion, Inc., Roanoke, VA]

Security robots need many sensors. Besides light and temperature, they need infrared or light sensors for people or animals. Detectors are needed for smoke and gas. Water detectors respond to flooding from a leaking roof or a broken water pipe. Sound detectors hear footsteps, breaking glass, or voices. Vibration detectors locate doors being forced or blown open. Ultrasound systems find unexpected obstacles, perhaps equipment or products being removed. The robot may also need radiation detectors.

The detectors must work together as a system, with one backing up another's finding. This is because the robot's controller should not make a judgment based on single occurrences. For instance, home smoke detectors sometimes go off while food is being cooked, after a certain concentration of smoke. It is easy to check immediately to be sure that no fire hazard is involved.

A robotic security system makes a sure judgment by having several types of sensor back each other up. For instance, if the noise and vibration go off at the same time, someone may be trying to break in. Vision, motion, and heat detectors may all have to be set off to indicate an intruder in the halls.

A higher heat level may mean that someone is in the room. But there is a difference between an unauthorized person and, for example, a mouse. The controller may be set to sound an alarm if the heat radiation is high enough to indicate a person, but not a mouse. It may also compare what the sensors have detected with past patterns. If conditions are above normal patterns, the computer signals building security or telephones for the police.

Security and surveillance are tasks that sophisticated robots can perform well. The U.S. military probably uses more of them than the commercial world does. The *2001 World Robotics Survey*, by the United Nations Economic Commission for Europe, in cooperation with the International Federation of Robotics, lists just 60 worldwide in 2000. However, it projects 1,800 by the year 2004.

Security robots are not just for offices and factories. At least one became available in 2000 for patrolling at home, ready to sound an alarm or call the police in case of intrusion. Spy-Cye, built and sold by personalrobots.com, is a package ($2,995) consisting of a core robotic unit, a spy camera, Ma-N-Zap mapping software, and a Cye T-shirt.

The nine-pound Spy-Cye is controlled through a graphic interface with any PC. Point-and-click controls allow the user to create a map of the premises by dragging an icon along the desired path. It has an on-board, 16-bit, 16-MHz controller that handles serial communications with the PC, motion control, dead-reckoning navigation, and obstacle detection sensors. Its batteries recharge at a home-base unit. It moves at three feet per second, can run on any floor surface, and can move between carpeted and bare floors. Its 12 servo-motors have a 38.3:1 gear ratio.

The U.S. Navy's Deep Drone ROV is lowered from the deck of the U.S.S. *Grapple* for an in-ocean check before beginning a search and recovery mission. [U.S. Navy photo]

Underwater Robots

The oceans cover two-thirds of the Earth's surface. They can be hostile and dangerous to humans. But the oceans contain forms of life, natural resources, and treasures from the past that people want. Until the 1950s, the ocean depths were visited only by a handful of scientists and adventurers, wearing heavy and clumsy old-fashioned diving suits or in hard-to-maneuver diving chambers.

Then, the ocean floor became an important source of petroleum. Oil companies needed to install and service huge offshore oil platforms and pipelines. New lightweight machines were developed to dive beneath the ocean's surface. Since robots were being developed at the same time, the two technologies were teamed up.

The military was one of the earliest users of undersea robots. The U.S. Navy used them to locate lost missiles. They even found an intact hydrogen bomb (H-bomb) lost from a U.S. warplane off the coast of Palomares, Spain, on January 19, 1966.

By the late 1970s more than 100 undersea vehicles were working in offshore oil fields across the world. Today there are too many to count. Besides the oil industry, they are used by marine biologists, archaeologists, mineral prospectors, and oceanographers. Environmental protection is another use. For example, robots have been used to investigate oil tanker accidents and discover how much oil has leaked from the ship and how much ecological damage has occurred. Even the tourist industry uses them to take visitors to underwater habitats.

Finding sunken treasure is another use for robots. They have been used to find sunken ships, ranging from 16th-century Spanish ships filled with gold to the ocean liner *Titanic*, which went down in 1912.

Undersea vehicles come in two major types. *Submersibles* are really small submarines. Some carry small crews or passengers groups. Larger vehicles are the underwater equivalent of tour buses, carrying as many as 48 passengers. *Remotely operated vehicles (ROVs)* have no crew or passengers. They are connected by communication lines to ships or platforms.

Not all underwater vehicles are robots, but many have such robotic features as manipulators with from three to seven degrees of freedom, specialized end effectors, feedback sensors, and vision systems.

THE SUBMERSIBLE *ALVIN*

Alvin is famous as the deep-ocean submersible that found both a lost H-bomb and the *Titanic*. It has been used (and upgraded) since 1964 for scientific research by the Woods Hole Oceanographic Institution in Massachusetts. One of *Alvin*'s most interesting scientific discoveries

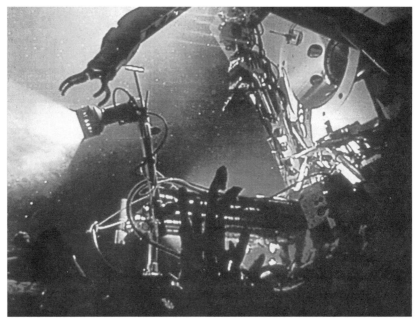

The submersible *Alvin,* **showing its robotic arm** (Rod Catanach/Woods Hole Oceanographic Institution)

was of colonies of tube worms and other organisms living near hot-water vents deep on the ocean floor.

The 18-ton vehicle is operated by a pilot and can carry two other people on three-day missions. *Alvin* can dive as deep as 13,000 feet—a depth at which no unprotected human diver could survive.

Alvin has two hydraulic manipulators that are controlled by the ship's pilot to lift and carry. One arm has six DOF and can lift about 200 pounds. The other arm has seven DOF and position feedback and can lift 250 pounds. For safety, the arms can be released from the inside and discarded, if necessary. Specialized end effectors can grasp or cut objects, take samples of the ocean floor, and place markers.

REMOTELY OPERATED VEHICLES (ROVS)

Remotely operated underwater vehicles—ROVs—have many functions. They can be used to dig a trench, take a soil sample, or lay and bury a communications cable or pipeline on the ocean floor. They can also inspect a ship hull or service an oil wellhead.

Most ROVs are controlled from ships or oil platforms through cables or *tethers*, which carry computer signals, pictures, and usually electric power. A few have their own power supplies but receive information through the tether. Some lack tethers and are controlled by TV signals. (ROVs of this type are sometimes called "autonomous," though they are not.)

Some ROVs are designed to swim in the water at specified depths. Others work on the ocean floor, either in a fixed position or mobile, able to crawl across the bottom on tracks.

In 1990, the AT&T Communications Company used bottom-crawling robots to lay fiber-optic cables across the Atlantic and Pacific Oceans and the Caribbean Sea. The cables are used for high-speed data transmission as well as voice phone calls.

One Japanese robot called MARCAS (*MA*rine *R*obot for *C*able work *A*ssistance and *S*urveillance, made by KDD Laboratories) can test soil samples, detect problems in buried communications cables, and dig the cables up and repair them. The robot has black-and-white and color vision systems, sonar, one seven-DOF and one five-DOF servo-controlled arm, and special cable grippers and cutters.

Another robot called Aquarobot (designed by the Japanese Ministry of Transport) can walk across the sea bottom on six spider-type legs at six feet per minute. It was designed and built in the late 1980s to monitor the construction of seawalls.

Deep Ocean Technology, Inc.'s Bandit robot services oil well-heads—the place on the ocean floor where the well meets the base of the platform structure. It inspects, cuts, and replaces parts using two six-DOF arms, force and touch sensor feedback, and a lighted TV system to view its work area. Bandit uses the platform structure as its pathway.

Jason and Medea are a two-part ROV system for multisensory imaging and sampling, operated by the Woods Hole Oceanographic Institution. The two are connected by cable. Jason has one six-DOF motorized manipulator that allows very precise control of position and rate of speed. Both vessels can operate to a maximum depth of more than 18,000 feet.

For sample taking, Jason's manipulator can lift more than 30 pounds. As samples are collected, they can be transferred automatically to Medea without stopping the collection process. Jason's cameras, lighting system, and sonar (sound) sensor systems allow precise mapping of the work area. The information is then automatically transferred to computers for storage.

The operator controls Jason by a remote *joystick*, watching its progress from Jason's four cameras. Both vehicles have automatic navigation systems, which improve their performance and reduce the operator's workload.

Hospitals, Medicine, and Genomics

Robots are playing an increasing role in medical treatment, medicine preparation, and research on the human genome.

SURGERY

Robots have assisted in surgery on both people and animals for many years. An early medical robot, invented in 1992 by a physician and a veterinarian, was a robotic arm called Robodoc. It was a modified

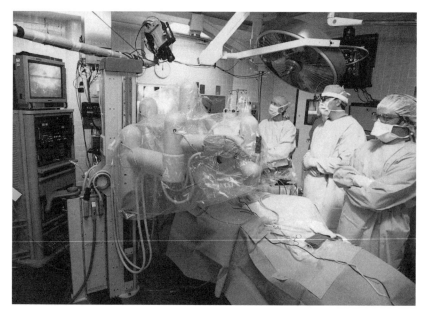

Dr. David Brown (left) and Dr. Robert Michler (center) observe the first minimally invasive heart surgery technique using robotic technology at the Ohio State University Medical Center. [Courtesy of Ohio State University Medical Center/ Jim Brown, photographer]

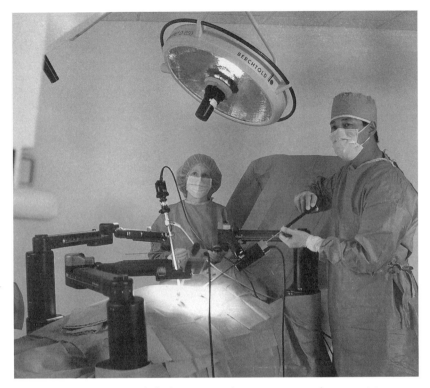

ZEUS robotic system at work during surgery [© Computer Motion/photograph by Robbi Bennett]

industrial robot, seven feet high and weighing 500 pounds. Robodoc was used in a Sacramento, California, hospital in several cases of brain surgery. First it measured the angle for entering the skull. Then the robotic arm, with a drill as a hand, cut through the bone. In a veterinary clinic, doctors used the robot to drill a hole in an elderly dog's leg bone so an artificial joint could be inserted. The patients recovered very well in all the operations. The surgeons said that the robots allowed much more precise drilling than with handheld methods.

In recent years, two new factors have made surgery on internal organs much safer for the patient and precise for the doctor—the use of tiny surgical incisions (called *minimally invasive surgery*) and very flexible robots that work under the surgeon's direction.

Before the surgical robots were developed, surgeons had to work with tools at the end of two-foot-long handles or *probes*. Then,

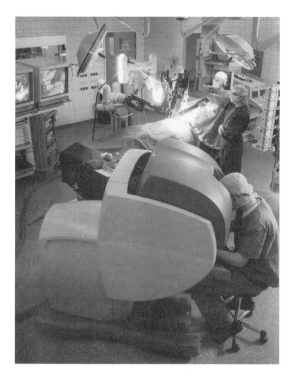

The first da Vinci™ Surgical System installation [© 2001 Intuitive Surgical, Inc.]

improvements in robotic design made the robotic hand just as flexible as a human one. Combining the two techniques has transformed delicate procedures like operating on a damaged heart.

For instance, with the da Vinci Computer-Enhanced Surgical System, designed by Intuitive Surgical, Inc., the surgeon works from a robotic workstation, not directly on the patient. The viewer shows a 3-D image, provided by several cameras working together. After cutting three small holes, the surgeon inserts two robotic arms, then directs their use of scalpels and clamps, as if holding the tools directly.

HELPING PEOPLE RECOVER FROM INJURIES

Robotic *exoskeletons*—skeletons on the outside of the body—have also been tried to help partially paralyzed people to walk. Wearing the computer-controlled exoskeletons like boots, people have been able to walk limited distances. Other experiments are working toward robotic artificial hands, arms, and legs. Robotic wheelchairs are also under development.

Genomics

In 2001, scientists finished decoding the human genome—the DNA "operating instructions" for each person's body. Robots played important roles in completing the work a year ahead of schedule.

The genome consists of long strands of a supermolecule called DNA, stored in the nucleus of each cell in the human body. Each tightly coiled DNA molecule is made of just four kinds of smaller molecules, called G, A, C, and T for short. Reading (or "sequencing") DNA is the process of discovering the order of the four smaller molecules. For instance, part of a sequence might be G G C A T A.

Preparing the DNA requires a series of chemical steps. First, other cellular materials must be dissolved. Then, the DNA must be uncoiled and heated to start a "digestion" process to cut it into sections. Next, it's sorted electrically by size and washed and dried. Finally, it is attached to a supporting stable material so it can be handled.

When the project began, in 1986, this work was done by hand or with simple machinery. The human DNA molecule is huge—more than 3 billion small molecules long. This meant that the laboratory workers had to perform the same tasks many times. The humans' robotic replacements greatly speeded up the work, reduced costs, and maintained consistent quality.

In the mid-1990s, there were no commercial robots that performed these tasks, so the Lawrence Berkeley National Laboratory in California, for instance, developed and built two specialized robots—DNA Prep and DNA Track. The DNA Prep robot, used for special processing, was equipped with laboratory equipment and materials and a pneumatic gripper. It moved on an overhead track, dispensing chemicals and preparations onto processing plates at preset locations. The controller ran on a personal computer.

The Prep Track robot was more complex and flexible. It contained several processing modules and more could be added at any time. One module retrieved the processing plates containing genetic materials. Then each plate was processed by a separate module, each with its own controller. There also was an overall system controller that determined the order in which individual modules of material were processed. Finally it stored the finished plates.

By 2001, commercial robots were available, and the laboratory replaced its "homemade" robots with several commercial proprietary systems, such as the Beckman Coulter, Inc., Biomek FX, which has more flexible plates, can handle more chemicals at the same time, and

Biomek FX robot used for genomic analysis [Beckman Coulter, Inc.]

works twice as fast as the original robots did. Other commercial systems include Matrix Technologies Corp.'s Platemate, a stationary robot with a single controller for adding liquid processing chemicals to the DNA, and the Robbins Hydra with a Zymark Corp. "Twister" (revolving joint) for "pick-and-place" work.

DNA's job is to direct the cell to produce proteins from which the body is built. Now that the DNA has been sequenced, the next step is identifying how each protein's structure is linked to the way it works in the body. Robots are involved here too. They purify the protein and turn it into stable structures. When this work is done, they link their results to computer databases that are immediately available to scientists throughout the world.

Robots at Work in the 21st Century

Robots are a large part of the work scene. But their roles will be even bigger when some longtime problems are solved. (As one expert has said, robotics is like ice-skating. It's harder than it looks.) One problem is that robots are not accurate enough. And they need greater intelligence.

For industrial robots, accuracy means getting the hand in just the right position. Accuracy involves the metal an arm is made of and the arm's speed. When metal moves, it vibrates. The faster the arm moves or the more it is extended, the more it vibrates. An arm cannot begin working until it stops vibrating. And vibration also disrupts the arm's path, so that it doesn't always end up on target. The problem is to move the arm quickly to the target without too much vibration.

The force and torque of the end effector at work can also affect accuracy; so can the softness or stiffness of the work material, friction at arm joints and as work is performed, and changes in temperature and humidity. Even gravity plays a role—the weight of the arm itself and the weight of what it is lifting.

Two ways engineers are trying to improve accuracy are by developing better control of motion and speed and better use of sensory feedback.

Robots also need to be "smarter." What does "intelligence" mean in robotics? The next chapter on the field of artificial intelligence will explore this question.

4

ARTIFICIAL INTELLIGENCE

Suppose you're sitting at a desk in a closed room. From time to time a conveyor belt delivers cards with Chinese characters written on them to your desk. You don't know how to read Chinese. But you have to take the cards and process them according to a list of rules that you must follow exactly. When you finish with each group of cards, you place it on a conveyor belt that takes it out of the room. Outside the room is the person who has sent the cards into the room and who receives them when you are done.

Before you processed them, the cards didn't make sense to the person outside. After you processed them, they do. Your processing provided just as much information as if you were fluent in Chinese, that is, as if you had read the cards and made sense out of them all by yourself. In fact, the person waiting outside the room can't tell the difference. Did you perform an intelligent act?

In this "Chinese room"—a famous scenario in the artificial intelligence field—a human has worked with the cards in the same way a computer works with data. People who believe that the person was acting intelligently in the Chinese room may also believe that a computer can act intelligently—that it can think.

In the past, the question "can computers think?" could be answered "yes" only in science fiction, with a character like Moxon's chess-playing robot. Or the "yes" could be wishful thinking, as the people who worked The Turk may have done.

Today, the same question is being seriously asked and answered by computer scientists. Some say yes; others say no. But this serious discussion is the foundation of a branch of science called artificial intelligence—AI, for short. And an intelligent computer could be the basis of an intelligent robot. Artificial intelligence grew out of the development of the computer in the 1940s. It was in 1950 that Alan Turing, one of the founders of AI, stated the basic test of an intelligent computer:

> If a computer can solve a problem that requires thought, and if a human expert cannot tell whether a computer or a human being solved the problem, then the computer was thinking when it solved the problem.

This "Turing test" is based on a simple idea, that thought is just moving symbols around, putting them together, and taking them apart. The idea can be stated the other way too: Moving symbols around is thought. That means the brain is a computer and nothing more. This idea is called *strong AI*, and it has many supporters. But it also has critics.

Other scientists believe that thought is more than combining, dividing, and rearranging symbols. They say that the Turing test is incomplete because thought includes understanding of the meaning of the symbols. They say that the Chinese room shows that processing symbols is not the same as intelligent thought.

A third group of scientists says that not enough is known about how humans think even to try building computers to do the same thing.

What Does the Human Brain Do?

If the human brain is a computer, it is a far more complex one than any built in a factory or laboratory. Combining and rearranging symbols is just part of the brain's work. It also has a way of storing information so that even the smallest clue of language or image can bring a whole idea, story, picture, or series of events into the person's mind. This is called *associative memory*.

It can use new knowledge to give meaning to something the person didn't understand before—the proverbial light bulb turning on or the "aha." The brain solves. It sets goals. And even in an uncertain or unexpected situation, it usually plans an action to get the person through it.

INTELLIGENCE

What does it mean to be intelligent? We humans define intelligence in terms of ourselves, so it is possible to define the term by seeing how a person works in the world. A normal person can understand the structure of living things, objects, and events—the way they are put together. A person understands how they relate to one another, and to himself or herself. A person knows how to change or rearrange them to make new relationships or even new things—to give them new meaning. Any person can make a plan to solve a problem, then carry it out, adjusting it along the way, if necessary. The same person can remember all this experience and knowledge, communicate it to others, and learn from it.

Each person is aware. The individual recognizes himself or herself and understands how everything else is "not self." A person knows what he or she knows and (usually) doesn't know and how to act in each case.

Some of these skills are built into the human brain. Others must be learned. For instance, using a language to communicate seems to be built into the human brain. But people learn to speak English, Spanish, or another language.

There are different types and levels of intelligence. Some species are more "intelligent" than others, in terms of controlling and using their environment. For instance, orangutans are skilled at making and using tools; dogs and cats are not.

Artificial intelligence means computer intelligence. There are two ways to develop AI: Program a computer to act like the brain or program it to provide intelligent results, even if the method used to get them isn't brainlike.

MIND AND MEMORY

Are a person's mind and memory the same as the brain? Mind and memory are very old ideas. They date back to the ancient Greeks, long before scientists began learning how the brain works.

The brain is a complex structure of cells *(neurons)*, pathways, and electrical and chemical signals. They work together to process and store information. Today many scientists believe that the term *mind* means how the brain works and *memory* is how it stores information. Memories are formed and kept in many parts of the cortex, the "smart" part of the brain. They are coordinated by another brain part called the *hippocampus*. But mind and memory are handy words, and even scientists continue to use them.

LEARNING

Memory is the storage of past problems that are solved, facts learned, sensory information, and emotions felt. Once someone learns how to ride a bicycle—balance on the two wheels, steer, lean into a curve—the brain retains the knowledge. People usually learn all these skills a few at a time. At each session, the brain stores what a person has learned. This means that next time the person can start where he or she left off without starting again from the beginning. The skill might be self-taught or learned from a teacher or demonstrator.

Furthermore, the brain doesn't store bike riding or other knowledge all by itself. A person can apply the knowledge learned about bike riding to other situations, such as riding a motor scooter or a motorcycle. It helps even when learning to drive a car. This type of learning is called *generalization*.

A computer or robot is intelligent if it can learn things, follow examples or instruction, and develop its own method of learning how to solve a problem. It must be able to "remember" (store) this new knowledge in a way that it can use in the future. Many AI researchers are trying to develop systems that can generalize from their experiences and stored knowledge.

Are There Intelligent Artificial Systems?

Does anything called "artificial intelligence" actually exist? So far, the results have been mixed. But there have been some successes, mainly in what are called *expert systems*.

EXPERT SYSTEMS

Almost everyone at some time has gone to the doctor after feeling unwell, such as having an abdominal pain. In the examining room, the doctor usually asks the patient a series of questions, like: Where does it hurt? Have you been exercising? Did you eat anything unusual? The doctor or nurse probably takes the patient's temperature and pulse and listens to the heart and lungs with a stethoscope. After a few more questions and some pressing on the abdomen, the doctor may decide to take a blood sample for testing. Once she has all the information, she compares it with what she learned in medical school and in her years of practicing medicine. She then rules out all but one possibility. Her diagnosis: a pulled muscle from playing volleyball. The patient will feel fine in a few days.

How did the doctor make the diagnosis? First of all, she needed knowledge of many different diseases and their symptoms. By asking questions, she learned about what the patient had done and eaten. She learned more about the person from the physical examination and the laboratory test. Once she had enough information, she used a series of rules to eliminate the most unlikely problems.

For instance, if the patient did not have a pain in the lower right abdomen and if the white blood cell count was not high, then she could decide that the patient probably didn't have appendicitis. If the temperature, breathing, and heartbeat were normal, she could decide to rule out other problems.

Expert systems or *knowledge-based systems* are computer programs that follow the same plan. Each one contains

- a body of knowledge,
- a method of getting information about a specific problem to be solved,
- a set of rules or other method to make use of the information, and
- a set of instructions for making decisions that solve the problem.

Why would anyone want an expert system to do what people are already so good at? The doctor in the example did just fine without a computer. But there are many situations where a physician can use help. For instance, a patient may have several medical problems that require treatment at the same time. Will medication for one problem interfere with the medication needed for another? What was the outcome for other patients in the same situation? Which combination of medicines works best? How have other doctors handled the situation? Are there two possible treatments? Which one would be better for this patient?

The doctor could put the basic information into an expert system. Then the computer could rule out unlikely diagnoses and treatments that probably won't work. This would let the doctor spend more time finding out about this patient and the best way to treat his illnesses.

BUILDING AN EXPERT SYSTEM

Medicine is just one field where expert systems can be used. Someone can construct an expert system that contains the skills of business, technology, or of everyday life. An example from everyday life is the expertise needed to wash a car, perhaps to pass on the knowledge to someone who has no experience.

The first step is to define the subject and purpose of the system. The next step is to find a *domain* (area) expert, learn how he/she works, and put the expertise and method into the computer. The step after that is to devise a method of reasoning.

The system also requires a program structure that holds the knowledge and reasoning method and lets the user pull out specific information and combine parts of it in different ways. The last element of the system is an *interface*, a way for someone else to use the computer and the expert system by asking for information or answers in plain English.

Domain Expert

An expert in a specific area has two types of knowledge. One type is the information itself. The other type is the method or way of using the information.

An expert at hand washing, or detailing, a car knows the specific equipment needed: a bucket of water, a sponge, some detergent, a hose, and a drying cloth. Another part of the expertise is the method of washing the car, perhaps starting with the top and working downward, saving the trim and glass for last. An expert may have a unique way of using tools, for instance, a way of holding the hose to remove dirt from the tire wells. But that's just general skill. An expert also knows how to deal with unusual problems, such as removing road tar, achieving a brilliant shine without harming the paint, touching up scratches, and removing or concealing rust spots.

The point is a general or quick description of accomplishing a goal isn't enough. The expert system must include the operations an expert may perform automatically, without even thinking about them. It may take careful questioning to learn some of the expert's methods. The system must also include adaptations of the expert's methods for unexpected conditions, such as how to wash a convertible.

Reasoning

Reasoning is the application of knowledge to present conditions and future plans. There are two major kinds of reasoning. One kind takes past experience and makes easy-to-use logical or practical rules from it. This is called *shallow*, or *heuristic*, *reasoning*. The other kind analyzes knowledge, experience, or a problem to find out its basic structure—what is called *deep reasoning*. Both kinds are used in expert systems.

SHALLOW, OR HEURISTIC, REASONING

Biologists and neuroscientists (scientists who study the brain and other parts of the nervous system) say that under stress, the body and brain are designed to help the person survive for approximately the next half hour. Survival skills include quick thinking, improvising, using a plan that worked on an earlier occasion, using common sense, and even acting first, thinking later.

Everyone knows these strategies. People frequently use them while taking a test, on the athletic field, at a party, or in a frightening situation.

Sometimes heuristic, or shallow, reasoning is put in the form of rules like this one, often used by baseball batters at the plate: "If the last pitch was a slow one, then this pitch will be a fastball." A person can use these if-then rules two ways for problem solving. One way is by working through a series of connected rules (a chain) until he or she reaches a conclusion or solves the problem—what is called *forward chaining*. Or the person can form a possible answer (a *hypothesis*) and work backward through the series of rules to see if they and the facts support it *(backward chaining)*.

People may want to use both methods to solve a problem. The person can work forward through the rules until it is possible to say "maybe the answer is so-and-so." Then he or she works back from the end to see if the facts support the answer. For example, after a few questions in the examining room, the doctor probably decided that "a pulled abdominal muscle" was a good possible diagnosis for the patient's situation. Then the doctor worked backward through a series of if-then rules to see if the blood test and abdomen examination supported the diagnosis.

Most early (or *first generation*) *expert systems* used heuristic reasoning. Medical diagnosis and treatment is very complicated so several systems were developed for medical use. One was Stanford University's ONCOCIN, which gave advice on how to manage cancer therapy. ONCOCIN was designed to ask for test results and other information about the patient. Then it took its knowledge about the disease and various treatments, plus a large set of rules, and developed a plan for the patient's treatment. It could also revise the plan as treatment progressed and new information was entered.

CADUCEUS was another early medical expert system, developed at the University of Pittsburgh. (It was named for the traditional symbol of medicine, showing wings and snakes wrapped around a staff.) It formed hypotheses based on more than 100,000 cause-and-effect rules, then tested them against known causes and effects of several hundred diseases and symptoms.

Other expert systems cover interpretation of irregular heartbeat, assessment of chest injuries, and the prospect for recovery from chronic liver diseases.

Medicine isn't the only use for expert systems. Several are available to assist geologists looking for the best place to drill for oil. The searcher can enter information about a location's soil, underground formation, and other features into an expert system. It will respond by estimating with how likely an oil strike is. Other expert systems are used in business, particularly in evaluating credit applications.

The military also uses expert systems. Those for combat situations are called *battle management.* An expert battle manager plans a military action and oversees the movement of troops, weapons, and other equipment. The manager also knows how ready the troops are for battle and manages strategy and even the firing of weapons. It receives information as a battle progresses, so that it can update all its plans and make changes to meet new conditions. Numerous battle management systems are part of computerized warfare plans in all branches of the military.

DEEP REASONING

In real life, someone who has used heuristic reasoning to get through a tough situation may go back over it in his or her mind. This will allow the person to figure out what worked and what did not. The person may compare the situation to previous ones and see what they had in common. Such comparisons may let the person determine how the situation might have been better handled. It may also provide expertise for successfully dealing with similar situations in the future.

This is deep reasoning. With it the person analyzes a problem's structure, how it works, and how its parts interact. Then he or she uses the analysis to make a model solution for the problem. The model can also be adapted to handle other problems. Expert systems like this are sometimes called *second generation expert systems.*

An example is ABEL, a medical system developed at MIT. It used shallow reasoning to connect test and examination results with diseases. But it also used deep reasoning to create models that showed how diseases cause the symptoms that people have.

Deep reasoning is more thorough than shallow reasoning, but it takes time. That means that it does not work well in fast-changing situations that must use lots of information, like robotics.

The Future of Expert Systems

Scientists have been working on expert systems for several decades. They have learned that "being an expert" is much more complicated than they first thought. An effective expert system must have appropriate structures and rules. The computer's power, speed, and memory size are important too. Also, everyday knowledge and common sense are much more important than scientists first thought.

Two expert systems, called Cyc and Soar, are trying to overcome these problems. Cyc has been under development by Cycorp, Inc., of Austin, Texas, for two decades. It knows at least 1 million facts. It has rules for mathematical facts as well as uncertainties and the current beliefs about them. Cyc's developers call it "formalized common knowledge." Its design and its contents allow the intelligent examining and organizing of unstructured information. It is useful for intelligent applications such as understanding speech, merging of databases, and e-mail handling, sorting, summarizing, and prioritizing. One version,

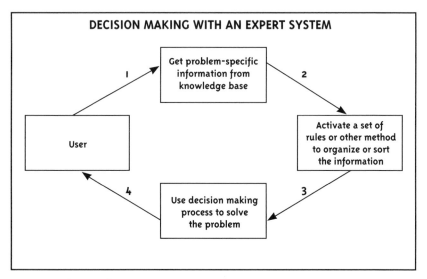

Decision Making with an Expert System   [Source: Ellen Thro original design]

OpenCyc, is available to the public without charge. The Department of Defense is using other versions.

Soar, developed by a nonprofit group, is an environment for the development of intelligent computer systems. Each decision it makes combines relevant knowledge, interpretation of sensory data, and memory created in previous problem solving.

Uncertainty

One of the main differences between the way a human thinks and the way a computer works is the kinds of questions to be dealt with. By design, today's computers handle yes-no and true-false questions such as "2 + 3 = 6. True or false?" Many situations can be put in this format, and rules can be written to handle them.

Real life doesn't always work this way. Students, for example, often deal with questions such as these: Is it going to snow today? Is my hair too long? Will I be able to get my homework done and still have time to watch television? In other words, many real-life decisions involve questions where the answer is "some" or "maybe," instead of "yes" or "no"—situations that involve *uncertainty*.

There are two kinds of uncertainty. One is *fuzziness*, or lack of precision. For instance, a sweater might be sort of red in color or slightly baggy. The other kind is *probability*—the *chance* that something will happen or the degree to which something will happen. People deal with both kinds of uncertainty every day.

FUZZINESS OR
LACK OF PRECISION

Fuzziness means that each number or other characteristic takes in a range, rather than single or "crisp" value. Instead of the ordinary (or *crisp*) equation $4 + 5 = 9$, you can make the numbers fuzzy. A fuzzy four is included in a range between three and five. Fuzzy five is in the range between four and six. The answer is a fuzzy nine, which takes in a range between seven and 11. The designer of a fuzzy system determines the actual range.

You can see fuzziness at work in a clothing store. Once most clothing came in standard numerical sizes—6, 8, 10, 12, 14, 16, and 18, for instance. The numerical sizes differed by about an inch in waist and chest measurements. Today most clothing is grouped in S, M, L, and XL sizes. Each one includes a much wider range of measurements. S, for instance, often includes the numerical sizes six and eight. L includes sizes 14 and 16. Now each measurement covers a fuzzier range of at least one and one-half inches.

This increased fuzziness gives the manufacturer and retailers several economic advantages. There are fewer sizes to order, make, and display, and the workers can produce the clothes faster because accuracy is less important.

Fuzziness also allows more stable, efficient equipment operation and more realistic decisions in real-world situations. Testing is an example.

Some standardized tests require "crisp" true-false (or yes-no) answers. In the real world, few situations require such "all or nothing" answers. For instance, if the speed limit is 50 mph, it's technically illegal to drive 51 miles per hour. But traffic and other road conditions may force you drive a "fuzzier" 50 mph—say, in a range between 48 and 53 mph.

Other standardized tests incorporate fuzziness, as well as personal emotion or belief. For instance, each test question might be in the form of a statement. The test taker is offered five choices—agree strongly, agree, neither agree or disagree, disagree, strongly disagree.

In 1965, the computer scientist Lotfi Zadeh developed the first fuzzy system for decision making. The first fuzzy equipment was built in the early 1970s, when a British engineer named Ebrahim Mamdani combined fuzziness and an expert system to control a commercial steam engine.

Fuzzy Consumer Products. Commercial equipment must work efficiently under changing conditions. For example, an ordinary washing machine with a "crisp" system typically offers three water levels, "low," "medium," and "high," and three temperatures "hot," "warm," and "cold." The agitation speeds might be "delicate," "normal," and "heavy duty." You choose the settings, perhaps by guesswork if your load is mixed—combining towels, sheets, and underwear, for example.

Now suppose you have a "fuzzy" washing machine system. Instead of the "crisp" settings, this machine's water levels might be "sort of low," "in between," and "pretty high." The temperature settings are "pretty hot," "lukewarm," and "pretty cool." The agitation speeds are "very fast," "mid-speed," and "quite gentle."

Like other commercial fuzzy-based equipment, this one works without much human help. It evaluates the load's characteristics during the first cycle, then automatically chooses the settings.

Fuzzy systems also run consumer equipment like vacuum cleaners and air conditioners. The fuzzy air conditioner is said to maintain a more uniform room temperature than a conventional one. It also saves more than 20 percent in energy costs.

Fuzzy systems are also the secret behind cameras that deliver sharp pictures even if the photographer has shaky hands. Cars sometimes use fuzzy systems as well. A Hyundai four-speed automatic transmission, for example, uses fuzziness for smooth shifting on hilly terrain, during aggressive driving, or in other uncertain conditions.

On the most practical level, such fuzzy systems are easier and cheaper to design and put into action than "crisp" designs that achieve the same end. The end products themselves look no different. No one can tell whether a vacuum cleaner or a car, for instance, has a fuzzy system or a crisp one.

Fuzzy Commercial Equipment. One of the first large commercial fuzzy systems was the subway in Sendai, Japan, developed in 1987. Fuzziness is also used to schedule banks of elevators in commercial buildings. An experimental fuzzy system was designed to control operations at a sewage pumping station in Shanghai, China. The system is

based on the relationship between the sewage's flow rate and the amount of storm water running into the sewers.

Other Fuzzy Systems. Fuzziness is also used in the business world. One system rates the quality of bonds based on characteristics of the company issuing them. Another determines creditworthiness of auto parts suppliers, also based on the companies' financial characteristics.

Probability

Suppose the morning weather report states that the day will be cloudy, with a 40 percent chance of snow. The view from the window shows a single fluffy white cloud in a blue sky. The question becomes what kind of clothes to wear. Does the person believe the weather forecast or think it's wrong? Perhaps the snow-laden clouds are just beyond the horizon.

Now suppose that the sky *does* show gray clouds, and the forecast is 40 percent chance of snow. This means that the forecaster thinks it's a little more likely that it won't snow than that it will. Still, weather changes quickly. The chance of snow might increase to 50 percent or greater during the next five or six hours. Or it might decrease to 30 percent or less. Again, do you believe the forecast? What type of clothing will you select?

Your decision will be based on several things. One factor is your past experience. Another is the opinion of a trusted friend or family member. Personal choice in clothing is still another factor.

People make decisions in a very uncertain situation like this all the time. A computer cannot do it even once.

Some expert systems have been equipped to handle a degree of certainty. Some do it with rules. Others use *certainty factors*, which give a different weight or importance to different kinds of fact and information. Sometimes the weight is how strongly the expert believes that the fact or information is true. For instance, the person in the previous example may believe that heavy dark clouds in the morning always mean snow by afternoon. So that belief could be twice as important to the decision maker as the weather forecaster's statement that the clouds will be blown out of the area by midmorning.

Of course, we can always update our decision making if we get later or better information. For instance, if a gray sky starts turning blue before you leave the house, you may put the heavy boots back in the closet. Scientists call this process *nonmonotonic reasoning*. As with true-false reasoning, the object is to work with what you believe is true. You then use that belief to solve the problem.

Computers for AI

The computers people use have slowed the development of intelligent systems. Even workstations—computers that handle large amounts of data very fast—process problems one at a time, that is, serially or sequentially. This is the same way the first computers worked more than a half century ago. AI needs more brainlike computers that are capable of handling multiple problems at the same time.

TALKING TO INTELLIGENT SYSTEMS

The most intelligent system in the world is valueless if people have trouble using it. The best way is to use English or some other human, or *natural*, language, rather than a computer language. That way the user and the computer can ask questions or display information as easily as two people talking on the phone or face-to-face.

Language is the main ability that separates humans from all other species. Neurological research taking place at many medical schools, using computers, CAT scans, and cooperative subjects, is creating a map showing which parts of the brain control its various functions, including the centers where language is controlled. Neurolinguistic research with such "brain mapping" at numerous universities, such as the University of Minnesota, Brown University, Howard University, and the University of California at Los Angeles, has provided much evidence that language is built into our brains. We use language to communicate facts, ideas, stories, and emotions. Communicating with another person is hard enough. Even though we've all been using language since we were two years old, we still have problems. For instance, "Did you make your bed?" has a very different meaning than "You did make your bed, didn't you?"

If you want to understand language, you must understand its structure (*syntax*) and its meaning (*semantics*). In addition to this basic understanding, you also have to summarize what you read or take a short statement and expand it. A person must be able to turn a statement into a question or use one word in place of another.

A computer program that uses natural language will need those abilities too, though at present it's much harder to use a natural language with a computer than with another person. Once you begin talking to another person, words or thoughts can be left out, usually without destroying the meaning. People will expect an intelligent computer program or robot to do the same thing.

Practical applications like voice commands for computers and translations between English and other languages are just the beginning. More humanlike versions are already showing up in expert systems and other areas of AI and robotics research.

Voice recognition systems have opened up computing and computer-related jobs for many people who are unable to use conventional keyboard/screen computers. Children with similar disabilities also benefit. Users include the visually impaired; people who are unable to use their hands because of birth defects, brain tumors, or strokes; mentally retarded people; and those with cerebral palsy. In addition, several federal laws and also many state laws mandate computers for the disabled in both schools and workplaces.

Intelligent or brainlike computers may also multitask, work within *grids* or *decentralized parallel processing*, run neural networks, and recognize patterns in human or environmental settings.

Multitasking or Parallel Processing

Suppose a computer could work on several problems at the same time or break a big or complex problem into sections and work on all of them at the same time. This parallel processing would be faster and more efficient for many problems. In fact, it would be much more brainlike.

Early in the 1990s, computer scientists thought that this would be done on single "supercomputers." These were very fast computers with thousands of parallel processors ("massively parallel processors") to solve very large and very complex problems, like advanced robotics, advanced graphics, and for other intelligent problems. One massively parallel computer used 64,000 processors working together.

Today, massively parallel computers are specific-purpose machines. In the United States, they are mainly used for weapons simulations. In

Japan, they analyze weather data from satellites and represent it graphically. In 2002, the fastest supercomputer is Japan's NEC Earth Simulator, which studies weather, climate, including global warming, and earthquake patterns. It has more than 5,000 processors and a top operating speed of 35 trillion calculations per second.

By the mid-1990s, general-purpose supercomputers had been replaced by two technologies. One was the faster and much cheaper work station that is now used for conventional graphics. The other was the Internet, whose popularity connected millions of ordinary people using ordinary computers. In 1996, a different model was created, called decentralized, or distributed parallel, processing.

David Gedye and Craig Kasnoff, astronomers at the University of California, Berkeley, were trying to discover intelligent life elsewhere in the universe (SETI). They had more data than they could afford to process. They decided to use home computers connected to the Internet. How?

Computers often are not actually processing information, even when the user is working at the machine. The SETI scientists asked for volunteers to process the data in their computers' "free" time. (The program is called SETI@home.) By 2000, the SETI scientists had half a million active volunteers and had accumulated half a million years of processing time.

Many other volunteer and commercial organizations have followed SETI's success. And distributed processing is taking a step forward with a concept called *grid computing*.

GRID COMPUTING

Grid computing moves beyond SETI@home by formalizing the speed, power, and software of computer systems on a regional basis, making them available to anyone. The computer grid, as it is called, follows the model of the electrical grid, which carries a steady supply of electricity from many sources to many consumers. The computer grid was first discussed in the 1950s. It is possible in the 21st century because today's computers, data storage systems, and other hardware and software are available and inexpensive. The ultimate goal is to have a worldwide computer grid.

A major initiative for a grid in the United States was announced by government research facilities in 2001. The hardware network is called the Distributed Terascale Facility (DTF) and relies on clusters of computers at Argonne National Laboratory, the San Diego Supercomputer

Center (University of California at San Diego), California Institute of Technology, and the University of Illinois at Urbana-Champaign. The software that will link the computer clusters is called TeraGrid. The facility will be built by the computer centers and such companies as IBM and Intel.

The grid is scheduled to go on-line in 2002. At its maximum performance, scheduled for 2003, it will be capable of performing 11 trillion calculations per second and store more than 450 trillion bytes of data. Once the grid is running, groups of scientists will compete for its use in projects such as data from distant galaxies, analysis of the human genome to find cures for diseases, and simulations of world climate change and biodiversity. The National Science Foundation will determine which projects will be chosen.

A grid that is linked more closely to commercial ventures, including Web services, has also been announced. Called the Globus Project, it will use open-source computer code and Web protocols, such as the XML coding language, to allow Web services over a grid. Microsoft, IBM, and several more specialized companies are supporting Globus.

AI also needs computer languages that can handle brainlike computing.

AI Languages

The computer languages that most people use aren't designed to handle most complex AI problems. Widely used languages like BASIC, PASCAL, FORTRAN, and C were designed to process numbers. But AI languages must be very flexible to handle a natural language, rules, and other structures. Two languages that are often used for AI are *LISP* and *Prolog*.

LISP

LISP (meaning *LISt Processing*) processes lists of words, groups of words, and other lists, which can be flexibly rearranged or changed. For instance, each of the following names is a list:

(Alicia)
(Michael)
(Bettina)
(David)

Each of these is a list that includes the original name and that person's brothers and sisters:

(Alicia (Marcia))
(Michael ())
(Bettina (Andrew, William, John))
(David (Nicholas))

You could search these lists for people with no brothers or sisters. One advantage of LISP is that a simple program can easily be expanded to a more complex one. If Alicia and David get married, the programmer could add definitions that would let you ask about the relationship of Marcia and Nicholas.

LISP is also useful in writing website descriptions in HTML. To speed things up, lists of macros (a single command for a series of operations) can be created the same way as other LISP lists are. LISP-based Yahoo!Store is a program to create storefronts on the website Yahoo!Shopping. It is thought to be the most popular e-commerce software, running as many as 14,000 stores.

PROLOG

Prolog means PROgramming in LOGic. It is used to write series of rules that are backward chained to prove propositions. Some AI developers like it because it can handle uncertainty. Suppose this database is in the computer:

birthday (anne, feb).
birthday (luis, apr).
birthday (kate, july).
birthday (paul, mar).

as in: Anne's birthday is in February.
You could then ask the computer questions, such as

?-birthday (anne, mar).

meaning "is Anne's birthday in March?" The computer would answer

no

If you ask

?-birthday (anne, x).

meaning "When is Anne's birthday?" The computer responds

x=feb

And the question

?-birthday (x,feb).

meaning "Who has a birthday in February?" is answered

x=anne

Prolog-based software has become widely used for project knowledge management in the construction industry. Contractors, architects, and engineers use it. Programs developed by Contractors Business Systems, for instance, allow project management, comparison of multiple project databases, and creation of Internet-based work teams.

Brainlike Neural Networks

The human brain's basic operation unit is a specialized cell called a *neuron*. Neurons are connected in networks, much like telephone networks. They send signals that combine electricity and chemicals along fibers called *axons* and receive signals along other fibers called *dendrites*. A signal is transferred from axon to dendrite at a point called a synapse. Neurons don't work alone. They are organized in various levels of operation.

Problem solving takes place within a level and also between levels. A problem is usually divided up and given to different teams of neurons in different parts of the brain—what is called parallel distributed processing.

The artificial neural network—usually called simply a *neural network* or *neural net*—is loosely based on the brain model. This is partly because scientists have not discovered exactly how the brain works. An artificial neural network is composed of paths and *nodes* (sometimes called neurons).

The shortcuts that people often take to walk across a field or park help explain how a neural network works. Often, a person starts by trying several alternate shortcut routes before deciding on a regular one—the fastest or easiest route to the destination. Once the best basic path is selected, the person uses it regularly. At first, the path may be faint, but each time it is used, it becomes a little more distinct. Perhaps other people start using it too, and soon all the grass is worn away. The best path, the one that gives the best results, becomes the most popular one.

A network may have a single level or several. Each level contains many neurons or nodes. If the network is given a number to identify, signals are first sent along all possible pathways between nodes. When processing is done, some of the paths lead to better answers than others do. They have done better at recognizing the number.

Then network training or learning begins. In the park story, being blacktopped may strengthen the best path. In a neural network, the pathways that lead to better answers are strengthened, so they will send more signals in the future. Paths that lead to incorrect answers are weakened so they send few or no signals. When the network is fully trained, only the pathways that lead to correct answers—correctly recognizing the number—are carrying signals. The next time that number is put into the network, the strengthened pathway will recognize it quickly.

Neural networks are good at pattern recognition, such as recognizing letters and numbers. Neural network computers are also used to recognize handwriting. Some experimental robots also use neural networks. In the real world, neural nets are used in such fields as pattern recognition in chemical analysis, medical image analysis, computer-aided tracking and characterization of homicides and sexual assaults, control systems, and brain-wave analysis.

Scientists are constantly learning new things about how the brain works. What does all the new information mean for robotics? The next chapter will explore intelligence in advanced robots.

5

INTELLIGENT AND ADVANCED ROBOTS

Robotics and AI are closely connected. For many years, researchers have wanted to use advanced, intelligent robotics as a special way to reach goals or solve problems. It is special for two reasons. First, robotics means physical action. A computer stays in one place and solves problems by manipulating electricity. A robot moves itself and other things. Second, people like their computers to work as fast as possible, but they do not usually put time limits on them. A robot's problem solving and actions must take place in *real time*—time that is measured on the clock or in our own bodies. A person at bat in a softball game sees the ball coming and hits it in real time.

Does intelligent robotics mean a robot that can take a person's place in the Chinese room or pass the Turing test? Some people say it does. If so, are there any intelligent robots?

Are There Intelligent Robots?

The first robot meant to be intelligent was Shakey, built in the late 1960s at the Stanford Research Institute. Shakey had its own on-board

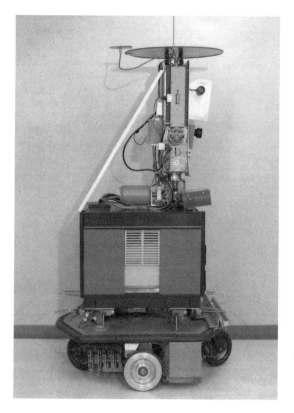

Shakey, the first intelligent robot, built in the 1960s [Printed by permission of SRI International, Menlo Park, CA]

computer for some operations, but a separate large computer performed most of them, sending commands by radio. Shakey could see, using a television camera. It had wheels and a rangefinder for navigating through its world with dead reckoning. Bumper detectors let it avoid hazards. It understood English language commands. It made plans and learned from experience. But Shakey worked in a very controlled and simplified world, not the everyday one. Even so, Shakey couldn't make quick decisions.

Since then, improvements have been slow. There are no truly intelligent robots yet. This is partly because there are no truly intelligent computers.

Better bodies are also required for intelligent robots. They will need more humanlike hands and better ways to move around. Teleoperation and sensors must be improved. And they will have to transmit more realistic information to their human operators or overseers—what is called *telepresence* or *haptics* (from the Greek word for "touch").

How Smart Should Robots Be?

How smart a robot should be depends on how you define intelligence. In the 21st century, there are several definitions. One follows AI research. Other scientists are defining intelligence more in terms of sensors—how much of the world the robot detects and how the controller uses the information. Still other scientists are examining a robot's behavior as it interacts with its world, rather than in terms of its brain power.

ADDING ARTIFICIAL INTELLIGENCE TO ROBOTS

According to AI research, before a robot can do any intelligent work, its brain needs the basic AI skills: expert knowledge and decision-making power, ready for use. The kinds of work robots are being designed to do may be surprising. Some jobs may seem simple, for instance, reading the utility meters in people's houses—the meters that measure how much gas, electricity, or water is used. To read the meter, someone just has to look at the dials and write down the numbers or draw lines on a diagram showing where the dial arrows or needles are pointing.

Being able to read the meter was one of the important intelligent skills of an advanced robot named Hermies, built at Oak Ridge National Laboratory. Hermies, actually a series of robots with that name, had several "advanced" features. It had two arms, one with seven DOF. It could navigate through its environment. And it had advanced sensors. It also had more computer power than most robots. It carried a parallel computer with 16 processors that let the robot move autonomously to its workplace. It could then quickly perform a complex task. It had additional computer power in a separate controller. The parallel computer had an expert system for navigation, with rules for decision making.

Isn't it a little silly to use all this robotic power to read a utility meter? Not at all. Reading the meter involves planning, sensor analysis, and complex control—all intelligent actions. First the robot must see the meter and get close enough to read it. It has to navigate carefully and park straight in front of the meter, so its reading will be accurate. Then its vision processor locates the needles, which it registers as lines.

Next it computes the needle angle on the dial. Finally it finds the angle on a table in its program that gives the correct reading.

An intelligent robot needs a plan to follow as it works. The plan should include goals along the way. It should also be able to deal with failures and—most of all—the unexpected. The robot also has to learn from its experiences. And it has to be aware of what it knows and doesn't know.

Robotic Planning

People make plans for their own actions all the time. Planning involves setting down each action that you must take to achieve a goal. You must know the amount of time each action requires. You also must decide the order in which each action is to be started and at what point in the plan each must be finished. In addition, you must figure out whether actions will overlap.

For instance, in the car-washing sequence, a wax may need time to dry before you can begin polishing. You can use the waiting time as a rest period, or you can use it on other washing activities, such as cleaning the wheel covers, wiping the windows, or vacuuming the inside. By the time those chores are done, you can return to polishing.

Robotic planning can take several forms. The controller can store a series of general plans, then adapt the one most suited to a specific situation. In another example, a person may know how to paint large surfaces with a roller and paint tray and how to use a brush to paint small areas. Each one is a general plan. The person can customize the roller-and-tray plan for specific projects, like painting a room wall.

Old plans that worked well can sometimes be adapted by revising a few elements. Suppose a sentry robot patrols the hallways of a company's building. When the company moves to a new building, someone can simply revise the plan's geographical features.

Another type of planning is the one people use in everyday life, for instance, when the toast burns during breakfast preparation or when a barricaded road forces people to take an alternative route. In these cases, a person starts with a plan but changes it on the spot as the action proceeds and conditions change or if the plan isn't working well. A robot's plan can use the same method.

Planning has another side too. Sometimes it includes understanding someone else's plan from his or her actions. Opponents in individual or team sports do this frequently. Diplomats, business executives, and generals must try to predict the other side's strategy and tactics. Then they can adapt their own plans or make new ones.

One way to analyze someone's action is by breaking it into its basic parts. This is like taking a machine apart to see how the pieces fit together. Another way is to try to understand why the actions are taking place—what purpose they serve. After that, you can predict actions that will also serve that purpose. Or you can write an if-then rule: If the light in room 128 is on at night, then it's all right for people to be moving around the area. Therefore, the sentry robot will not send an alarm signal.

Knowing the order in which events take place is very important in planning. Does the robot's arm begin moving toward the assembly line before the part arrives, after it arrives, or just as it arrives?

PLANNING FOR AUTOMATED FACTORIES

Planning skills will be very important when robots work with other planned and programmed systems in totally automated factories. Computers and robots will perform all the operations, including ordering materials, producing parts, assembling them, and preparing them for shipment.

Assembly and other operations will be coordinated with the help of an AI planner. This is an intelligent computer system that sets goals and priorities. It takes into account the product wanted, the numbers to be produced, and the type of equipment and materials needed. It definitely includes the way assembly robots interact with their work environment. Robots with several arms will be at work. Teams of robots will work together. Many operations will be going on at the same time. Each will have its own plan, but all the plans will have to work together too.

AI PLANNING FOR BUYING ROBOTS

How does a company shop for an industrial robot? Most do it by comparing features on various models, talking to salespeople, and observing what models other companies are buying and using. It's the same process an individual uses to buy a bicycle or an automobile. Like any major piece of equipment, a robot is expensive. The buyer wants to be sure it will do its work and fit into the overall manufacturing plan. The decision will be even harder to make for advanced, intelligent robots. Where can companies turn for help?

AI may provide the answer. Scientists in Germany, the Czech Republic, and elsewhere are experimenting with expert systems that decide which robot is best for the buyer's needs.

Using AI this way expands the idea of AI planning for factory production. A German expert system starts by asking the purchaser: Do you really need robots for the tasks you want done? If so, what characteristics do you want for your robot?

The expert system also helps the purchaser choose among the many models for sale by comparing each model's features with the tasks that he or she wants done. To answer these questions, the system builds the knowledge base. The German system contains 200 plans for using robots in various ways. It also holds all the specifications of the robots on the market. And it analyzes 500 automated tasks.

The specific information that the expert needs about the buyer's factory is also entered, such as the tasks to be done and kinds of material to be worked on. For instance, a question on the screen asks:

What motion systems do you want? (choose one)

- electrical

- pneumatic

- hydraulic

The system also asks questions about the type of grippers needed and what tasks will be performed. The buyer can also draw a diagram of the robot's work area showing its size, layout, where the robot will be placed, and what other equipment will be there.

Now the system goes to work by comparing each task, material, and other job characteristics with every robot on the market. Each variation is a robotic world. The worlds are compared to the one the buyer has created. If no robot is exactly right for the buyer, the system makes trade-offs among robot features. Finally, it comes up with the "best buy" and explains its decision.

Another expert system, by Czech scientists, helps design a robotic assembly line. The program is written in the AI language Prolog. It forms rules from information about available robots and the tasks to be done. Then it uses the rules to make its choice.

Most planning for intelligent robotics will be about the robot's tasks. Then the plans must be transferred into arm and hand motions. This is an advanced form of teaching and learning.

Interacting with the Environment

How do you teach a robot to give a sheep a haircut? It's hard enough to teach a robot to do assembly-line work. Teaching one to work in the outside is even more complicated. Shearing a sheep is like giving the animal a skinhead haircut, only all over its body. A human shearer has to hang on to a squirming animal, then use electric clippers to quickly take off all the wool, without injuring either of them in the process. Doing the job well requires great skill.

Some Australian robotics scientists took on the job of teaching a robot to shear a sheep this way. First, they used computer graphics to design a generic 3-D sheep. Then they gave it the measurements and shape of a specific animal. They added the shape changes caused to breathing and other natural sheep movements. An expert sheep shearer's hand and body movements were added in. When everything worked on the computer, the scientists programmed the shearing motions into a robotic arm and clipper hand.

Introspection

The final AI skill an intelligent robot needs is introspection, the ability to examine its own reasoning methods. This lets the robot tell the difference between what it knows and what it doesn't know. Introspection will let an autonomous robot react faster and better in a changing environment, enabling it to decide whether to act on its best knowledge or wait until it has more or better information.

Improved Robotic Body Parts

Every robotics scientist and engineer would like to have fully autonomous robots at work in the real world. This dream is slowly starting to come true. Robots are showing increased intelligence by their use of advanced teleoperation and sensing and by putting their human operators on the scene with a "you are there" technique called telepresence or haptics.

For robots, better intelligence requires better bodies. This includes improved hands. The hands (end effectors) should be *dexterous*. That is, they should have several jointed fingers and an opposable thumb—the outstanding structural feature of the human hand and one that makes humans distinct from the great apes. Flexible cables serve as tendons to bend and straighten the fingers.

Mobility must be improved too. Robotic wheelchairs, forklifts, and carts are all being tested with new types of wheels. They can move, turn, and go over obstacles at the same time. The purpose is to improve dead reckoning navigation over rough surfaces. For instance, if the wheels can roll over an obstacle instead of going around it, there's less chance of getting lost.

Some walking robots have been equipped with touch sensors on their legs. They work like insects' feelers. They can reach out and touch the ground to know where to step next.

ADVANCED TELEOPERATION

Teleoperated remote vehicles and programmed robots are being employed for specific tasks. Future systems must put the two technologies together. This way, they will make better use of human operators, expert systems, and intelligent machines.

This will involve:

- Real-time expert systems to control on-line robots as they perform their tasks in, for example, hazardous environments
- Improved vision and force sensing, so that the controller can use the sensors' information for making decisions while a task is in progress
- Force-reflecting servomanipulators that are more reliable and easier to keep working accurately and reliably

Another way of defining a robot's intelligence is how much of the environment its sensors detect.

ADVANCED SENSING

Walking down a familiar hallway is a highly predictable process, requiring routine skills. People usually don't notice where they are, except to keep from bumping into someone. Today's sentry robots are skilled enough for patrolling predictable areas. But the detection skills they provide are not thorough enough for many situations where the highly unusual can occur: military posts, high-tech industrial plants, or other high-security areas. An intelligent system also verifies the findings and assesses the situation.

High-security robots need intelligent skills. One such skill is environmental awareness, interpreting many signals provided by humanlike sensors. Another is good navigation, the ability to move accurately through the building. The third is making fast and correct decisions.

Robart III is an advanced mobile sentry robot developed by Commander H. R. Everett and others at the Naval Command, Control, and Ocean Surveillance Center in San Diego. It shows how a controller can take information from many sensors and use it to move intelligently in the real world. Unlike Robart II, its predecessor, Robart III is a military robot. It can also respond to threats with a nonlethal dart system as it patrols its programmed area or one that a remote human operator sends it to investigate.

(a) Robart II, an advanced autonomous sentry robot developed by the U.S. Navy
(b) Robart III, enhanced with weapons capability [Space & Naval Warfare Systems Center, San Diego, Point Loma, San Diego]

BRAIN

Robart's intelligence comes from 13 computers in the robot's body that control its actions and take in real-time sensory information. The computers also monitor the robot's own operation and report abnormalities to the controller.

An ordinary personal computer performs the intellectual work as controller. This means planning navigation routes, keeping the two-dimensional "world" map up to date, and telling Robart how to avoid obstacles. With this system, the robot moves through ordinary offices and hallways and avoids known obstacles. When it confronts unexpected chairs and wastebaskets, it plans new routes around them. When Robart returns to its starting place, it automatically avoids the new obstacles found on the outbound trip.

The controller computer also evaluates the security situation. Robart has several kinds of sensors, 128 in all. Some are designed to detect intruders or other security problems. Other sensors help Robart navigate.

ENVIRONMENTAL SENSORS

Robart's environmental sensors tell the controller whether building conditions are normal or not. They measure temperature, relative humidity, barometric pressure, light and noise levels, toxic gas, smoke, and fire.

The robot also takes its own temperature to tell if the system is working normally.

NAVIGATION SYSTEM

Robart uses dead reckoning, its world map, and sensor information to navigate through the territory. A stereo vision system lets it see where it is going and compare the view with the map stored in the planner. Twenty-four range-finding sonar detectors circle Robart's swiveling "head." They help the robot keep track of its location as it travels.

Its bumpers contain touch sensors that tell which part of the bumper has collided with something. Each sensor has a tiny switch that sends a signal to the controller, which decides how to move past the obstacle. Infrared sensors located near Robart's wheels detect the floor. As long as it can "see" the floor, Robart won't fall down stairs or tip over crossing a dangerously high floor threshold.

This information goes to the controller, which tries to plan a new route. If it cannot, it uses voice synthesis to call for human help.

INTRUSION SENSORS

The environmental and navigational sensors and the navigation system let Robart perform its intelligent sentry duty. It has a whole different set of sensors to help: Six different types of sensor detect human intruders and infrared (heat) sensors detect body heat. Sensors that register differences in light level can record this.

Ultrasound (sonar) can detect human motion inside a building just as it can a submarine's movement under water. Microwaves, similar to radio waves, also detect motion by changes in the wave's size and shape. These waves can go through walls, the way a radio's incoming waves do, and detect motion on the other side. A video camera detects motion by comparing a current picture of the scene with a picture of the way the scene should be, stored in the computer.

Other sensors monitor vibrations, which might be made by a moving person. A sound system picks up noises.

If there is a possible threat, having many sensors is important. The microwave motion detector might pick up something that the other motion detectors didn't. The robot's sound detector may have heard something at the same time, but the building alarms may or may not have gone off. The controller takes all the sensor information and compares it with information stored in its memory about "normal" and "alarm" conditions. Then it makes a decision.

These multiple sensors also make it harder for intruders to beat the system. It takes more time and skill to disable or fool several different kinds of sensor than a single one. Once Robart detects an intruder, it will follow it and speak a command to stop. Its video camera provides a picture of the situation to human security guards.

Threat Response

During the last century's cold war, the military developed teleoperated vehicles armed with lethal weapons to be used in battles. Now, the military places much greater emphasis on law enforcement and operations other than war, such as peacekeeping and antiterrorism operations. With this in mind, Robart III was designed to include a nonlethal weapon system to protect itself, to prevent acts of sabotage or terrorism, and to detain hostile individuals or small groups until human reinforcements can arrive.

Robart's builders had several choices of weapon, such as explosively deployed sticky nets and chemical agents. They chose a gun with six rotating barrels (like a 19th-century Gatling gun) that fires tranquilizing darts. Activated by explosive bursts of compressed air, it can be programmed to shoot from one to six high-speed darts in 1.5 seconds. This operation can be performed automatically or manually. If a remote human operator is in command, the robot's aim is controlled by the video system. The weapon's accuracy is tracked by a visible red laser or, for covert operations, by a near infrared laser.

To simplify control for a human operator, the weapon system is tied to the robot's camera and steering systems. The controller uses information from its sensors to automatically screen out false signals and track the target before firing.

Telepresence or Haptics

Ordinary tools are fine for some situations, but there's nothing like being on the scene and touching the object. The sensors and manipulators on today's tethered robots—such as undersea explorers—may give the operator considerable control over robot operations. This, as you have already learned, is called teleoperation.

Telepresence or haptics transfers the robot's sensor experience to the operator. It can show:

- **Position.** People commonly reach far back into a dark but well-known closet for something on the floor. Even without looking, a person can judge location. The touch of a clothes hanger may signal the person to reach farther back and down.
- **Texture and rigidity.** Someone reaching for a tennis ball knows just what it should feel like: small, round, slightly rigid, and fuzzy. On touching a shoe or a volleyball, the person knows instantly that it isn't the tennis ball.
- **Size and weight.** People quickly learn to judge the weight and content of an object to be moved, as when carrying a trash can. By touching it and moving it just a bit, someone can tell whether it's full and heavy or only partly full and much lighter. This sensory information tells how much strength the person needs to pick it up and carry it.

- **Force and torque.** When beginning to remove the cap from a jar, the feel of the cap, plus past experience, tells the person how much twisting motion (torque) to use.
- **Environmental conditions.** Even without touching anything, a person's skin sensors can feel heat and cold and wind.

For a telerobot to do these jobs, telepresence could let the human operator experience them just as vividly, even from very far away. The key elements of telepresence are sensors, manipulators, hands, and vision systems that are almost as good as human ones.

EQUIPMENT FOR TELEPRESENCE

What equipment does a robot need to provide telepresence? According to some experts, it should have at least two arms and hands that work together. The arms need seven DOF with force control. They should be strong enough to lift, hold, and move large objects. The hands should be able to grasp things. They need force indicators and controllers. To see what the robot sees, the operator needs a head-mounted vision display.

USING TELEPRESENCE

What do you do with all this information? Most research is for use in the military and in space. Exploring the ocean floor and deep beneath the Earth's surface are other possibilities. All these environments are dangerous and uncertain.

Telepresence will be vital in unknown places where conditions can change at any moment. Suppose the robot is exploring underground geologic formations. This means many uncertainties for the human operator in deciding how to interact with the formations. If the robot lifts a particular rock, for example, will the surface cave in under it? So, the robot's sensory information should also be fed into a computer program that can predict what will happen after the operator acts. Based on the sensors' readings, the computer can simulate the operator's action. Then it can decide "it is safe to lift the rock and carry it ten feet away" or "if the robot picks up the rock, the ground will cave in under it."

Telepresence will also be used with autonomous robots. Even though they will be working on their own, they will send real-time sensory feedback to human supervisors.

PEOPLE-TO-PEOPLE TELEPRESENCE

Another way of using telepresence is in the long distance people-to-people encounters—for instance, in robotic surgery and other medical fields. Doctors often consult over long distances by telephone or on the Internet. When a doctor examines you in person, he or she may thump you on the back or examine an injured ankle carefully with the fingers of both hands. This low-tech procedure often gives the physician a very good idea of your condition, even before the blood samples are drawn or analyzed. Or a student learning to insert a needle correctly could benefit from a haptic training session on a dummy before actually working on a patient.

As surgical procedures become roboticized, they also distance the doctor from the patient. Telepresence is a way to restore the surgeon's full experience and provide a richer picture of the patient's condition.

A drawback to haptic systems in medicine is the technology. Most telepresence is based on force feedback sensors. But people generally sense an object directly through its texture, temperature, shape, vibration information, and even sound cues.

Haptic systems won't become more realistic until additional sensors can provide all this information.

ENHANCING THE HAPTIC EXPERIENCE

Stanford University's CyberGlove/Grasp is designed to improve the haptic experience by letting the operator control a dexterous robotic hand intuitively—just as he or she would if grasping the object directly.

One dexterous robotic hand is Stanford's Dexter, a two-fingered hand with two DOF in each finger and a motor at each DOF. Using a motor that lands the robotic finger smoothly on target, the way a human finger would, increases realism. Dexter's fingertips are strong enough to pick up an object that weighs about half a pound, such as a softball. The hand can position itself within less than 0.004 inch and can manipulate spherical, cubic, and cylindrical shapes.

HAPTIC FEEDBACK IN THE FIELD

G. R. Vermeij is a paleontologist at the University of California, Davis. He is blind and uses his hands—the equivalent of haptic feedback—to create mental pictures of his research findings. His tools and methods show the kinds of human sensors that are required for a realistic picture.

Dr. Vermeij focuses on examining objects that are one cubic foot or less in size. His hands explore the material in linear fashion—like reading line after line of Braille. Larger objects make it too hard to build the information into a picture of the whole object, such as a piece of coral reef. He uses these sensory tools:

- whole hand—to perceive an object's temperature and weight
- fingernails—to evaluate the surface texture
- lightweight stiff stylus (sharp stick) or hypodermic needle—to determine the surface's two-dimensional features. For instance, he drags a stylus over a shell surface to detect slight or deep ridges and to count features that rise above the surface. This method has been so successful that his sighted colleagues also use it.
- fingernail scrapes—destructive testing of surface hardness

To make use of his fingers' fine tactile senses, Dr. Vermeij never wears full gloves. Even in the coldest water, he wears only fingerless gloves. He "reads" an object with both hands, to assess large and small scale information at the same time. This is important in quickly building up a mental image that combines fine features and large shapes.

Sounds also add to the image. Walking across scallop shells sounds like walking on bone china. Dead coral underfoot sounds different than living coral.

Social Robots

Most robots have been designed for industrial work or, more recently, as toys. Robots as aids for the disabled, handicapped, and those undergoing rehabilitative treatment have lagged behind. Only in the 21st century are intelligent robots becoming widely available for these purposes.

TYPES OF SOCIAL ROBOTS

In general, there are three types of social robots:

1. Those with programmed movements in a structured environment, like HANDY 1. This robot, developed by Rehab Robot-

ics Ltd., in the United Kingdom, has a robotic arm and a series of module trays for eating, cleansing, and applying makeup. The user activates the robot's programmed movements by focusing eye movements on a switchboard at the back of the tray. This system is appropriate for people who can't use their hands and fingers, but it is hard to reprogram and doesn't add greatly to the user's autonomy.

2. Those with user-controlled movements in an unstructured environment, such as MANUS, which was developed at MIT. This robotic arm has six DOF and a two-fingered gripper. It can be customized for the person who uses it, but requires someone with use of his or her hands and fingers.

3. Those that combine programmed and user-controlled movements, such as FRIEND, developed at the University of Bremen, in Germany. FRIEND combines a MANUS robotic arm, an electrically powered wheelchair, and a personal computer. It can be controlled by speech commands, and the computer screen shows whether commands are understood.

The robotic architecture consists of:

- a **person-machine interface,**
- a **command interpreter,** which in turn connects to
- the **robot controller,**
- a **database** of programmed actions that is also connected to the controller, and
- a **feedback system.**

The robotic arm follows the preprogrammed commands automatically. It can also be trained to follow user-controlled sequences of commands.

OTHER HELPFUL ROBOTS

Several helping robots are commercially available or are nearing availability.

- **iBOT.** iBOT, manufactured by Johnson & Johnson, is a versatile wheelchair that overcomes the limitations of most ordinary wheelchairs. It has the usual two large wheels in back and two small ones in front. But it also has four medium-sized wheels that swivel, for instance, for climbing stairs or going over a curb. It can also raise its seat if the user wants to take something from a high shelf. And it travels easily across rough terrain, such as an unpaved trail or a beach.

- **MAid.** Robots usually work best in predictable and closed environments. The University of Ulm, Germany, has developed a wheelchair called MAid (short for Mobility Aid) to provide severely disabled people with the fine motor control needed in dynamic situations. For instance, rush-hour traffic in a commuter station is continuously changing and the corridors are crowded and narrow. To move through this, MAid incorporates a navigational system, speech-recognition system, touch screen maps, and decision-making software. These will make it much easier for the chair's users to take part in ordinary activities.

- **Robotic-like Extremities.** Computer-controlled prosthetic legs and hands, similar to those on robots, are also available. The Otto Bock C-Leg (Otto Bock Orthopedic Industry) is a microprocessor-controlled hydraulic knee, leg, and foot system that allows amputees who weigh 220 pounds or less to walk with a natural gait. It immediately responds to changes in walking speed and can be adjusted to an individual's unique walking patterns. An onboard computer collects data from strain gauges in the shin and a position sensor above the knee, then analyzes the wearer's gait 50 times per second. The system uses what Otto Bock calls "stance phase control" to allow natural movement on either flat or uneven terrain. For example, when a wearer places the natural foot lower than the artificial one, the knee's hydraulic pressure is reduced, allowing the knee to bend and the leg to swing forward. A lithium-ion battery in the knee can run for more than a day without recharging. The system costs approximately $40,000, almost three times that of an ordinary prosthetic leg.

An intelligent hand developed by Dr. Peter Kyberd at Southampton University, in England, has two fingers and a thumb that can grasp and handle objects, such as an egg, naturally. It weighs about a pound and can be operated by tensing two muscles in the forearm, producing small electrical signals. Electrodes in each muscle amplify the signals and feed them into the hand. Small microphones in each fingertip detect vibrations from a slipping object and adjust the amount of force needed to keep hold of it. The adjusting mechanism can also be turned off with a muscle twitch, if necessary.

Some scientists have taken an even more sensor-oriented approach to designing robots. Theirs interact intelligently without any brains (controllers) at all.

Brainless, But Intelligent Robots

The purpose of robots has always been to do jobs that people ordinarily do, whether on the spot or remotely. They expand and extend our physical abilities. Intelligent robots use human skills as they work—planning, setting goals, and achieving them. But are plans, reasoning, and goals really the way people interact with the world? Or do we usually follow some well-established behavior patterns that let us sense the world, then react and interact with it as it changes? Suppose "intelligence" is the way we really interact with the world.

A familiar example of depending on the senses to understand our surroundings is a Halloween "haunted house." It is an environment meant to trick our senses and trigger fear, even though people know it is all in fun. The haunted house is mostly dark, so that the senses of touch and sound must do most of the work. Reaching for a wall and feeling a curtain or cobweb or tripping over an unknown object makes it hard for you to form a mind-picture of what you're experiencing. Finally, by moving cautiously and following other sensory clues, such as a breeze or a ray of light, you can successfully complete the trip through the "house." In other words, people use their intelligence by reacting to the environment and successfully moving through it.

RODNEY BROOKS AND BEHAVIOR BASED ROBOTICS

Some robot scientists believe that these behavior patterns can be the basis for intelligent robots. One such scientist is Rodney Brooks at Massachusetts Institute of Technology, who has developed robots that work on the principle of behavior patterns. Genghis is one of his robots. Genghis is small, about 14 inches long. But like its "conqueror of the world" namesake Genghis Khan, it is powerful. When the six-legged robot walks around or climbs over a book, its behavior patterns are at work.

A robot like this doesn't start out "smart." Brooks's robots start simply by sensing the environment and end by taking action. If the robot is moving and it senses an obstacle, it moves around it. When this behavior works well, a slightly more advanced behavior can be added.

Someone in a haunted house moves through it more intelligently as he or she senses things and reacts to them. The robots work in roughly the same way. If everything goes well with the robot's simpler types of behavior, layers of even more advanced behavior can be added. First Genghis stands up. When it accomplishes this, it can also walk. If it is successful, it can then start climbing over books. Eventually, Genghis, or one of its successors, will have even "smarter" behavior patterns. For instance, it will create a map of its world as it explores and use it to avoid running into things.

Genghis works without having a central controller. Instead, it uses a "subsumption architecture," in which each behavior "layer" is controlled independently. But the layers communicate, so that no advanced behavior can start unless the simpler ones are up and running, just as a baby cannot start walking until he or she stands up.

Genghis's sensors tell it about its world as it moves on its six legs, each with two DOF. The sensors detect its pitch (motion up and down) and roll (as in rolling over or perhaps falling sideways off a book). It also has heat and force sensors.

According to Brooks, studying a robot's behavior in the real world provides information a person can't get from studying a more complex robot in a controlled world. He compares it to studying animal behavior in the wild, rather than in a laboratory setting.

HELPMATE, A COMMERCIAL SUBSUMPTION-BASED ROBOT

Subsumption architecture has had a commercial success in a hospital courier robot called HelpMate (produced by Pyxis Corporation, in San Diego). It is completely autonomous, battery operated, and designed to carry 6 cubic feet or 200 pounds of cargo, such as hospital records or medicines.

It uses a five-layered architecture and navigates with a radarlike pulsed laser beam (lidar) and directional range scanner. It doesn't need a track to follow and can go down corridors or from floor to floor, including getting on and off elevators unassisted.

HelpMate is in use in more than 150 hospitals in Canada, Europe, Japan, and the United States.

Microrobots

As small as Genghis is, other working robots are even smaller. Some are only about one-fifth the size of the period at the end of this sentence. Others are so small that they are invisible to the human eye.

Such *microrobots* are being developed for medical treatment. One works as a pump in an artificial pancreas to treat diabetes. A process called micromachining is producing tiny motors and sensors 0.004 inch in diameter that are made of silicon, the same material used to make computer chips.

Many microrobots are actually close relatives of the computer chip. Many microdevices are made in much the same way that computer chips are (though other techniques are also used, and other materials are also being studied). Computer chips are wafers of the element silicon on which integrated circuits are etched.

The process has been used to make pressure sensors so small that three can fit on the head of a pin. Microrobots are possible because the world of the very small is different from the world we know.

One such difference involves the direct conversion of electrical energy into mechanical energy, what is called the *piezoelectric effect*. This effect is too weak to be useful in normal-scale activities, including normal-sized robot actions. But it can be extremely useful in the world of the very small. One six-legged microrobot uses it to move individual biological cells in a laboratory disk, making it a useful future tool for biomedical research and medical treatment.

Micromechanics studies the movement and behavior of tiny machines as small as 0.004 inch in diameter. Very small lightweight devices behave differently—relating to their environment differently—than large heavy ones do. For instance, an ant can carry a large piece of dirt. A bug can walk across the surface of a pond.

An ant has a tiny *mass* (the measure of its ability to accelerate when a force acts on it). So the forces of inertia (the tendency to stay in one place or move in a straight line) and gravitation (attraction between masses) are much less than on an elephant or a person. Several scientists have pointed this out: If you reduce something's volume 100 times, you decrease its mass 1,000,000 times. Inertia and gravitation decrease by the same amount.

Some forces are more important for bodies with a smaller mass than they are for bodies with a large mass. Electrostatic attraction—the force that makes dust stick to a mirror—decreases only 10,000 times when volume is reduced 100 times. This makes electrostatic force much greater than gravitation—on objects with a small mass. Also, electromagnetic forces are much less on tiny objects than on large ones. This means that electrostatic forces are more important than electromagnetic forces. Another way large and small masses behave differently involves surface tension, which increases with the area of a surface. All of these physical laws are why a tiny bug can walk across the surface of a large body of water.

Friction and wear as parts rub against each other are different also, making microdevices the object of study of the discipline called *tribology* (in Greek, *tribos* means rubbing). This is important for several reasons. Scientists estimate that one-third of all the energy in the world goes to waste as friction. Friction cuts the efficiency of robots and other machines by reducing their output of useful energy, information, or communications signal. This is because when machine parts rub against each other, their physical and chemical properties change. They stretch or become hot or fail to work the way they're supposed to. Also, some of the material is worn off and thrown into the environment as waste and possibly pollution. Research is now under way to find out how friction and wear affect microrobots.

Newer technology is making smaller robots practical to build. For instance, a Swiss company, RMB Miniature Bearings, makes motors for autofocus cameras. They are so small that a ladybug looms over one of them. Scientists at Sandia National Laboratory are using one in a tiny research robot. With an 8 kilobyte memory and treads for motion, the robot can be programmed to follow a simple route at a rate of less than a yard a minute and avoid a hostile barrier. Parts cost only about $500, making multiple robots possible on a small budget.

Some day, microrobots designed for the microworld may be used routinely in intelligent valves and pumps implanted in the human body. They could help people with physical disabilities. They could even be sent to specific body locations, perhaps through genetic engineering techniques.

Even microrobots may seem big, compared to the next generation. These are *nanorobots*, 1,000 times smaller.

Nanorobots

Fantastic Voyage is a classic movie about a medical team that is miniaturized and sent into the bloodstream of a dying man to save his life. The movie, released in 1966, was called science fiction, and no one expected that this level of miniaturization would ever be possible. Miniaturization hasn't gone that far yet. But nanoscale materials are already being used as components in laser filters for telecommunications equipment and films that make windows self-cleaning, to name just two examples. Scientists are creating nanorobots that are small enough to insert into a human cell, to treat cancer and other serious diseases.

Nano (from the Greek word for "dwarf") means one-billionth or 10^{-9} of a meter or 0.00000004 in., so the nanoscale world of robots and living things is invisible to the human eye. The idea of creating useful machines on such a tiny scale was first suggested by the physicist Richard Feynman in the 1950s. The driving force behind nanotechnology since the 1980s has been technologist K. Eric Drexler. He believes that in the future nanoconcepts and equipment will cause important changes from health care to agriculture.

In 2000, President Bill Clinton announced the National Nanotechnology Initiative to encourage socially useful, but long-term research. For example, the Environmental Protection Agency (EPA) is sponsoring research to create nanotools that will remove very small contaminants from waterways. Other nanotools will continuously monitor the results.

Various states have also begun similar initiatives. The largest one is the California Nanosystems Institute to be run by two branches of the University of California.

EXPONENTIAL ASSEMBLY

To many scientists, one of the most exciting features of nanotools is the degree of precision possible. In the large-scale tools being made and used now, placement of the actual molecules or atoms isn't important. In a nanorobot that will enter a living human cell to treat cancer, for instance, the atom or molecule of medicine must be precisely positioned. If it isn't, the treatment won't work.

Positioning might be done with a scanning tunneling microscope, which works automatically in a three-dimensional, nanoscale world like a robot to precisely position particles and assemble systems.

Vast numbers of any tool will be needed to work in the natural-scale world, such as a waterway. This means that fast and inexpensive ways to make the needed tools must be invented. One method, suggested by Zyvex Corporation in Richardson, Texas, is to use a two-step process. First small robotic arms will be built. Then they'll be used to assemble the nanotools themselves.

Such "exponential assembly" begins with a single robotic arm with two DOF and a gripper and carefully laid-out parts. This arm then uses the parts to create a mirror image of itself on a facing surface. Next, each of the two arms creates a mirror image robot, and so on.

Other scientists hope to send the parts into a living cell, then assemble themselves into a working robot there.

These activities are expected to take a decade or much longer before they result in useful tools. But a major breakthrough came in 2000, when scientists at Cornell University created the first nanomachine, a "biomolecular" motor with a propeller attached. (Nanomachines are also called MEMS, for microelectromechanical systems.) When fueled by ATP (adenosine triphosphate, the molecule that supplies energy for chemical reactions inside living cells), the propeller rotated about eight times per second.

NANOCIRCUITS

Ultrasmall electronic systems may allow nanorobots to perform more complex work by turning two structures called nanotubes (tube-shaped carbon molecules) into transistors. Scientists at IBM first attached a pair of nanotubes to gold electrodes, letting them pass unchangeable electrical signals. Then the scientists covered part of one nanotube with potassium molecules, allowing it to change its voltage. This let the two nanotubes work together as a *Boolean logic gate* (or system) to answer true-false questions. This ability is the basis for computer information searches and complex programming.

Eventually, scientists expect to place thousands of carbon nanotube transistors in the same space now occupied by a single silicon transistor.

Of all the places where advanced intelligent robots will be used, outer space is one of the most exciting. The next chapter looks at the place of robots in space.

6

SPACE ROBOTS—PAST, PRESENT, AND FUTURE

Space begins where the Earth's atmosphere ends and the planet's gravitational effect is small enough to let objects remain in sustained orbits. It's less than 200 miles away—the distance between New York City and Boston. But it was only in the 1970s that people developed the technology to go there and stay for long periods of time in permanent structures.

Why are we drawn to space? Besides the desire to explore the unknown, scientists already know that space outposts could provide mineral resources, energy, and even new homes for people.

What's holding us back? There are two big factors, cost and technology. The cost of manned spaceflight is so great that no one country can afford it. Fortunately, robotics is a lower-cost alternative.

Robots and robotic technology have been our partners in space since the beginning. Cost and the hazards of working and living in space are the two main reasons robotics is playing an increasing role in the 21st-century space program. In space as on Earth, robotics is most valuable in three types of situations: where human life could be endangered, where the use of humans is very expensive, and where repetitive tasks requiring great precision will be performed. In space, robots are being used for exploration, scientific investigation, and construction.

Are space robots different from earthbound ones? Earth's atmosphere protects us from extremes of heat, cold, and radiation. Working in space can be hazardous for machines as well as people. Weightlessness doesn't affect robots the way it does humans, but temperature and radiation do.

Space robots must be made of materials that can withstand temperatures ranging from 400°F in the sunlight to over 200° below zero in the shade. Gamma rays (electromagnetic radiation) can also damage both materials and electronic circuitry. Another hazard is damage from flying objects. These can be meteorites or pieces of space junk left over from previous missions.

Humans working in space must wear heavy suits to protect them from all these hazards. This slows down their work rate. Scientists have decided that it is better for well-designed robots to work under these hazardous conditions instead.

Astronauts on the space shuttles use a teleoperated robotic arm to launch communications satellites into orbit. Robotic, science-oriented space vehicles have gone far beyond Earth's orbit. Spurred by the 20th-century's cold war, the principal space-faring nations were the United States and Russia (then the Soviet Union). Both countries have sent missions to our neighboring planets, Mars and Venus. Other ships are exploring the deep space that's beyond our solar system.

In the 1970s, both Russia and the United States put space stations into orbit. The final Soviet/Russian station was *Mir*, which in the late 1990s played host to astronauts from the United States.

In the 21st century, scientists in many countries are developing space technology by building and using the *International Space Station*. The main modules are contributed by the United States and Russia. Canada, Europe, and Japan are providing specialized modules and tools.

Space Stations in the 1970s

Beginning in the late 1970s, the Soviet Union launched a series of seven self-contained Salyut space stations. However, they could not receive supplies, fuel, or new crew members.

SALYUT

Salyut was an overall name for the Soviet Union's first seven space stations, covering both its four peaceful-purpose stations and three in the military-oriented Almaz program.

Military Salyuts. The military stations—*Salyut 2, 3,* and *5*—carried reconnaissance equipment to study military installations on Earth. *Salyut 2* depressurized after 13 days and was unusable.

Salyut 3 was successful in orbit. About 35 feet long and 13 feet at maximum diameter, it had an interior volume of 135 cubic feet and solar panels. Reconnaissance was carried out with a television system and a telescope, whose exposed film was sent back to Earth in a recovery vehicle. The station also carried out some medical and physical experiments. In 1975, its orbit decayed, and it burned up in the Earth's atmosphere on January 24.

Salyut 5, launched on June 22, 1976, was similar to *Salyut 3,* but carried out its reconnaissance with a wide-angle camera, infrared viewer, and other instrumentation. Film was developed on board and sent back to Earth by reentry capsules. Three two-man crews staffed the station. One stayed aboard for 49 days. A second flight, in October 1976, was unable to dock and returned to Earth, landing in Kazakhstan during a blizzard. When rescuers finally reached the crew a day later, they were surprised to find both cosmonauts (Russian astronauts) still alive.

The third flight placed two cosmonauts on board for 18 days, but the mission was cut short because of the illness of one of them. A fourth flight was planned, but the station's orbit decayed before the launch date. *Salyut 5* burned up in the Earth's atmosphere on August 8, 1977.

Peaceful-purpose Salyuts. *Salyut 1* was launched on April 19, 1971, and had two pressurized areas, one for the crew and another for equipment. Its purposes were to study the effects on the human body of long spaceflights and to photograph Earth from space. Other studies involved star spectra and cosmic ray studies and analysis of plant growth. Its gamma ray telescope was unable to get rid of its cover, however, making the instrument useless.

The first three-person crew visited the station on April 22, but their ship was unable to dock and they returned to Earth. The next crew docked on June 7 and remained on the station for 24 days. After they left for their return to Earth, their ship depressurized rapidly and they all died.

Salyut 1 completed 362 orbits of the Earth before it reentered the atmosphere and burned up over the Pacific Ocean in October 1971.

Salyut 4, launched on December 26, 1974, was similar to *Salyut 1*, but added three moveable solar panels. During *Salyut 4*'s two-month scientific mission, its three two-person crews conducted experiments with an X-ray telescope, optical sensors, and X-ray detectors. In addition, a flight of the robotic *Soyuz 20* spacecraft was successfully carried out. Launching on November 17, 1975, it spent three months docked at the space station, where its "crew" of plants and turtles took part in a joint experiment with the station crew. It then returned to Earth on February 16, 1976.

Salyut 6, launched on September 29, 1977, and the similar *Salyut 7*, launched on April 19, 1982, were more sophisticated stations, with two docking ports each. Improvements in *Salyut 7*'s station design and comforts—stoves, a refrigerator, and hot water—allowed crews to stay aboard for months, the record being 236 days in 1984. The station was also host to the second female cosmonaut, Svetlana Savitskaya, who was the first woman to take part in a space walk. Another visitor was the first British astronaut, Helen Sharman.

This configuration allowed simultaneous docking of the original spacecraft and a ship carrying supplies and new crew members. Under the Soviet Union's Intercosmos program, guest astronauts from other countries joined the resident crew aboard the stations.

Salyut 6 burned up in the atmosphere on June 29, 1982. *Salyut 7* was in orbit for about nine years, before it, too, burned up. The two stations' basic configurations and peaceful mission strongly influenced the design of the *Mir* station.

SKYLAB

The United States's first space station, *Skylab*, launched on May 14, 1973, was a multimodule station intended to orbit the Earth for many years. It was 84 feet long and a maximum of 22 feet wide and weighed almost 83 tons. More than 300 experiments were performed during the 172 hours that crews were on board. It achieved its main purposes of showing that humans could live and work during prolonged stays in space and that astronomy from space provided information that was impossible to obtain from Earth.

Skylab's principal parts were

- an instrument unit used during launch and initial operations to deploy equipment, the Apollo Telescope Mount, a solar observatory,

- an external laboratory whose film had to be retrieved and replaced by an astronaut,
- an orbital workshop, crew quarters with provisions for three astronauts for 84 days,
- a multiple docking adapter, for servicing ships, and
- an air lock module, which housed controls and instruments.

Its main purpose was to show that people could survive and work in a space-station environment for extended periods. During 1973 and 1974, three, three-person crews of astronauts stayed on the station, one for 84 days. Work performed aboard the station included biomedical studies of the effects of weightlessness, photographing Earth's volcanoes and earthquake fault lines, astronomical studies such as those of the Comet Kahoutek, and the effects of microgravity on processing of materials.

Skylab's air lock module let them work outside of the station—what is called *EVA (extravehicular activity)* or a space walk. *Skylab* stayed in orbit until July 11, 1979, when it was brought back to Earth, disintegrating over the Indian Ocean and western Australia.

The experience of running these stations gave both nations the expertise to design later generations of stations, with increasing use of robotics.

MIR

The first modular, expandable space station was the Soviet Union's *Mir* (Peace), whose core module was launched in 1988. Designed to orbit 225 miles above the Earth for five years, *Mir* was expanded and updated throughout the 1990s and continued to operate until 2001, when it was brought out of orbit and crashed in the Pacific Ocean.

Despite the weightlessness of space, *Mir*'s core module was designed with a floor-to-ceiling orientation, much like a boat's. This orientation was very different from the three-dimensional one used by the United States. *Mir*'s operations area had a green-carpeted "floor," light green "walls," and a white ceiling with fluorescent lights. The pastel-colored living quarters had a communal kitchen with table and range. Behind this were individual cabins—complete with porthole—for each crew member. The toilet, sink, and shower were in the rear.

The 19-foot long Kvant-1 was the first module added to the basic core. It was designed for astrophysics research, such as measuring electromagnetic spectra and X-ray emissions of galaxies, quasars (distant starlike bodies that emit large amounts of radiation), and neutron stars.

Two modules added later provided sensing instrumentation that could be adapted for robotic use. The Kvant-2 module had sensors for Earth science research. It also provided a self-contained maneuvering unit for a cosmonaut working outside the station on, for example, construction and electronic components.

Mir's Priroda module, added in 1996, carried a highly automated digital camera with data collection instruments called MOMS-2 (the Modular Optoelectronic Multispectral/Stereo Scanner), developed by the German Space Agency. MOMS-2 sensed the visible light and near-infrared light spectrums to detect erosion risk, natural hazards, and other environmental characteristics. An earlier version of MOMS had been tested on three previous spaceflights. Future versions will be part of more fully automated sensing instrumentation.

The 40-foot-long Kristall module was devoted to biological research, Earth observation data, and the effects of space on electronics and construction materials. Spektr, equipped with four solar arrays and other scientific equipment, was used to study particles in low-Earth orbit. The final addition was a U.S.-style docking module that allowed NASA's space shuttles to dock.

Mir was serviced regularly by automated Progress-M ships that brought food and other supplies to the station and removed used equipment and trash.

United States Space Shuttle

The U.S. space shuttle program has used a robotic arm (called Canadarm or the Shuttle Remote Manipulator System) since 1981. This arm, designed for heavy lifting and construction work, was designed and built by MD Robotics for Canada's National Research Council. Canadarm robotic arms fly on all space shuttles and have assisted in retrieving and deploying satellites and in delivering materials to the *International Space Station.*

Canadarm has six DOF—two in the shoulder joint, one in the elbow joint, and three wrist joints. It is 45 feet long, weighs almost half a ton, and can handle 585,000 pounds in space. The arm is mounted just outside the shuttle's payload bay—the doorway where equipment is moved in and out. It has been used on shuttle flights to retrieve and deliver satellites. It can also serve as a platform or transportation for an astronaut working outside. Canadarm can also perform more delicate operations, such as knocking ice from a clogged wastewater vent to avoid a possible hazard on reentry into Earth's atmosphere. It also has pushed a faulty antenna into the correct position and activated a satellite that didn't go into the proper orbit. In 1998, it assisted in the first assembly of *International Space Station* modules, the United States's Unity and the Russian Zarya (Sunrise).

The arm is controlled remotely by a shuttle crew member. To make contact, the end effector must grasp a sticklike grapple fixture on the object. The operator uses two controllers. The right hand uses a joystick to move the rotational hand controller on three axes. A trigger on top of it signals the end effector to capture or release the object.

The operator's left hand works a translational controller that moves the arm along its current axis. The operator looks out the shuttle window to monitor the arm's progress.

The shuttle's robotic arm is also used in construction of the *International Space Station*.

International Space Station

In the 1990s, it became clear that an advanced technology space station would cost more than any one country could afford. A planned U.S. space station became an international project, led by the United States and Russia. The U.S.'s contribution is based on *Skylab* and Russia's on *Mir*.

The *International Space Station*, which orbits the Earth at altitudes between 241 and 250 miles, is an assemblage of modules connected by docking ports and passageways. These were built on Earth and put into orbit for final connection. A framework of trusses strengthens the station, much like the supports on a large Earth-based building.

In style and technology, the station has two sections, the Russian and the American. Each section has its own interior design, docking mechanisms, and even the astronauts' space suits. Each section also has its own robotic tools. The crews, though, are mixed, with replacements commanded alternately by a Russian and an American. They also cooperate on some experiments.

The *Endeavour* shuttle's robotic arm performed the first robotic work—connecting the two original space station modules—Zarya, the

The first two *International Space Station* modules in orbit, the U.S. Unity (left) and the Russian Zarya (right), photographed by an astronaut on the Space Shuttle *Atlantis.* (NASA)

first module, was Russian, launched by Proton rocket to the station site in November 2000. The automated module provides station control equipment. Its solar array and batteries produce three kilowatts of electrical power. The next month, the *Endeavour* carried a U.S. module, called Unity, to the site. Unity is a docking station for future modules and also has a connecting passage to Zarya.

The connection procedure began when the shuttle captured Unity with the robotic arm and docked with it. The shuttle then captured

Zarya, also with the robotic arm, and connected it to Unity. When this was completed, astronauts performed several space walks to connect power and data cables.

In shuttle space station transfers, the shuttle's robotic arm is linked to the station's Space Vision System. This lets the remote operator monitor the transfer on a TV screen, instead of looking out the window. The system analyzes the screen image by measuring the object's

U.S. astronaut James S. Voss handles the main boom of a newly delivered crane, preparing to install it on the station. [NASA]

U.S. astronaut Scott E. Parazynski, with Canadian astronaut Chris A. Hadfield (not seen), handles cabling for installation of Canadarm2. (NASA)

location, altitude, and moving rate in real time. The operator then makes fine adjustments.

ROBOTICS IN THE U.S. SECTION

The *International Space Station*'s U.S. section will have a set of robotic tools called the Mobile Servicing System. This consists of

- a versatile seven-DOF robotic arm called Canadarm2, built by MD Robotics for the Canadian Space Agency,
- the Base System, a truck that weighs two tons and can carry a 23-ton load. It is designed to move along rails on the outside of the station, for instance, as it moves the arm from one location to another, and
- a 15-jointed Special Purpose Dexterous Manipulator, also called the "Canada Hand." This system will maintain and service the U.S. section after construction is complete.

Canadarm2. In 2001, the U.S. section's own robotic arm, Canadarm2, was brought on board from a shuttle. The arm is an advanced version of the shuttle's Canadarm. It is 58 feet long fully extended and weighs more than 1.5 tons on Earth. Its seven DOF include three in the shoulder, one in the elbow, and three in the wrist. They give the arm a greater range of motion than a human arm has.

Unlike the space shuttles' Canadarms, Canadarm2 is not installed on a static base. Instead, it can move end over end to one of the station's Power Data Grapple Fixtures. The ends of the arm are equipped with Latching End Effectors, which connect to the grapple fixtures. They provide the arm with power, data, and video. The arm can run autonomously or under an astronaut's control.

Other advanced features include force sensors that provide a sense of touch. The vision system consists of four color cameras—two on

The "Canada Hand" under construction in the laboratory [Courtesy of Canadian Space Agency www.space.gc.ca]

the elbow and two on the Latching End Effectors. The arm also has an automatic collision avoidance system.

Canadarm2 can be reconfigured or repaired in space. Since it is modular, any one of its sections can be removed or replaced on the station.

"Canada Hand" or Special Purpose Dexterous Manipulator. Canada is also providing this robot. It has two arms and can be attached to the end of Canadarm2. It will be controlled by station crew members using a robotic work station. It is capable of handling the delicate assembly tasks previously handled by astronauts during space walks. It can also conduct intricate maintenance tasks. It is designed to extend to 11.5 feet long and has 15 DOF.

ROBOTICS IN THE RUSSIAN SECTION

The Russian segment of the station will also have a set of robotic tools, including a robotic arm, an automated transfer vehicle, and two automated supply ships.

The European Robotic Arm. This is designed for attachment to the station's Science and Power Platform. It will install the platform's solar arrays and also help assemble and maintain the segment. The arm is a cooperative venture of the European and Russian space agencies.

It has seven DOF, works in three axes, and can operate with any six joints at a time. It is about 35 feet long and weighs almost 1,400 pounds. It is functionally symmetrical, with an end effector at each end. Each end effector will work either as a hand or as a base for the arm's operation.

This configuration, plus a camera on the end effector, lets the arm find various base points on the platform and precisely relocate itself.

Japanese Remote Manipulator System. The Japanese Experiment Module Remote Manipulator System is a robot with two different arms—the Main Arm and the Small Fine Arm. The system is designed to work on experiments on the outside of the Japanese Experiment Module and also for maintenance work. Each arm has six DOF and can move like a human arm. The Main Arm is more than 30 feet long and can handle about 15,000 pounds of payload. Its two long sections are connected at an "elbow" joint. A "wrist" joint connects to the end effector/grapple fixture.

The Small Fine Arm is about six feet long and can handle almost 700 pounds of payload. It has two shoulder joints, one elbow joint, three wrist joints, and a torque sensor and camera between the wrist joint and the end effector.

The arms also have force/torque sensors that "feel" the object, letting the arm automatically control its grip. The system has a 10-year life, so it is designed to be repaired or replaced in space from inside or outside the station.

An astronaut will monitor the arms' movements on a TV screen and manipulate both arms from a console inside the module. The console contains translation and rotation hand controllers, a laptop workstation, a TV monitor, a camera control panel, an arm control computer, and electronics.

Progress-M Space Ship. In the early phases of the *International Space Station*, an automated Progress-M ship, the same model that supplied *Mir*, will resupply the Russian section. Later, an Automated Transfer Vehicle will supply the segment.

Automated Transfer Vehicle. This vehicle, built by the European Space Agency, combines a spacecraft and a cargo carrier. It will be launched by an Ariane rocket from French Guiana. The vehicle is larger and more versatile than a Progress-M. It has a pressurized cargo module and can carry 8.3 tons of dry and liquid cargoes on the same mission. It is designed to deliver cargo, transfer propellant to the Zarya module, and control the station's altitude to maintain a stable orbit (called *reboost*).

It can dock with the station automatically or manually by the station's crew. Once docked, it can use the station's power system for up to six months during refueling and cargo transfer. The vehicle is scheduled to make at least eight trips to the station between 2003 and 2013.

OTHER ROBOTICS ON THE STATION

The AERCam is a free-flying robotic camera that may be used to inspect the outside of the station, including its acre of solar panels. A prototype of AERCam was tested on a space shuttle mission in 1997.

Robotics plays an even greater role in missions to Mars and Venus, because no manned missions are planned for the foreseeable future.

Missions to Mars

Mars has always fascinated people. The purpose of any expedition to Mars is to learn as much as possible about the Red Planet—so-called because much of its sand-swept surface looks red from Earth. Since ancient times the planet has been named for various gods of war or death. The name Mars comes from the ancient Roman war god.

In the 21st century, the United States has sent several successful robotic missions into Mars's orbit and has plans for more to orbit Mars or land on its surface. As with robots on Earth, there's no all-purpose Mars robot. Each is built for specific tasks in a specific part of the Martian environment. These determine, for instance, how autonomous the system will be and the types of sensors it will need.

First Robotic Explorers

Just getting to Mars has been difficult, even in the 21st century. The first actual spaceflight toward Mars was in 1962—the U.S.S.R.'s *Mars 1*, which was lost on the journey. Scientists say that only 30 percent of all flights headed toward Mars have reached the planet's orbit with major systems ready to function.

Mars is one and one-half times as far from the Sun as Earth is. It takes about 23 Earth months to complete its elliptical orbit. Its distance from the Earth varies between 35 and 250 million miles, depending on where the two planets are in their orbits. A spacecraft's trip will be somewhere in between. It will need a spiraling course to get there. This is because the spacecraft must escape Earth's gravity and avoid the Sun and other planets. So the actual journey from Earth could be 100 million miles long.

Mars is about half the size of the Earth. Its mass is one-tenth that of our planet, and its density is three-fourths that of Earth. Each Martian "day"—a complete rotation on its axis—is called a *sol* and takes about 24½ hours. Because of Mars's smaller size, its gravity is much less than Earth's.

> *Mariner 4.* Between 1965 and 1969, several U.S. and U.S.S.R. spacecraft flew by Mars or orbited the planet. The U.S.'s *Mariner 4* orbiting spacecraft—called a *probe*—in 1965 was the first to show conclusively that Mars had no canals, as a 19th-century astronomer had claimed. *Mariner 4* also discovered that the Mar-

tian atmosphere was composed mainly of carbon dioxide and that the terrain was much like the Moon's.

Mars 3. The first soft landing on Mars was by the Soviet *Mars 3* orbiter/lander, which launched on May 29, 1971, and arrived on the Mars surface on December 2, 1971. However, the lander's instrumentation failed after relaying 20 seconds' worth of video data to the orbiter. The orbiter itself made measurements of surface temperature and the atmosphere's composition until August 1972.

Viking 1 **and** *Viking 2.* The partly robotic American *Viking 1* and *Viking 2* spacecraft, which landed on Mars in 1975, provided all the information we have from the planet's surface. Each Viking was composed of an orbiter and a lander. The orbiters took pictures of the entire Martian surface and transmitted them to Earth, where the landing site was selected. The landers then separated and parachuted to the surface, where they took pictures and transmitted them back to Earth. They also took surface samples and analyzed them for composition and signs of life and studied the atmosphere and weather. The *Viking 2* lander also deployed seismometers, which measure earthquake activity.

The Viking ships sent back photographs and data until the early 1980s, informing us that surface conditions are very different on Mars and Earth. Mars's atmosphere—mostly carbon dioxide—is much thinner than Earth's. This means that, on Mars, the Sun's warmth dissipates much faster than on Earth. Mars is a very cold, dry desert, with temperatures near -100°F (a little colder than Antarctica) and no surface water. The Martian polar caps are a combination of ordinary water ice and dry ice (frozen carbon dioxide).

SOJOURNER ROVER

Sojourner was the first robotic roving vehicle to land on Mars, on July 4, 1997, as part of the *Mars Pathfinder* spacecraft. Pathfinder's software allowed it to land autonomously after a signal was sent from Earth. The landing procedure included opening a parachute to slow its speed and displaying a metal tape so the lander could slide slowly down to the surface.

Sojourner was primarily a demonstration instrument to determine if the design would work in the Martian environment. Its robotic features included

- **autonomous navigation,** including avoiding obstacles and hazards,
- **sensing and monitoring its environment,**
- **planning its actions,**
- **carrying out its assignments,** and
- **changing its behavior to meet changing conditions.**

The Sojourner Mars rover, demonstrating its abilities at NASA's Kennedy Space Center, Florida [NASA]

During the three months it operated on Mars, Sojourner showed that it could move small rocks, climb over them, avoid collisions with them, and perform other tasks in the Martian environment.

Sojourner stood almost 11 inches high, two feet long, and 18 inches wide. It weighed about 25 pounds and carried about 13 pounds of telecommunications equipment, computer, stereo camera, solar panel, spectrometers, and an autonomous steering system. Each of its six wheels had independent gears and was independently controlled for navigating through the soft Martian sand. It navigated by dead reckoning and kept track of its movements.

Commands were sent from Earth and stored in the lander, then requested by Sojourner. The rover had preset stopping points and sent position signals to the lander before moving to the next point. Its Earth station reviewed the photo images at the end of each sol and used them to plot Sojourner's commands for the next sol's work.

The name *Sojourner* (meaning "traveler") was for Sojourner Truth, the 19th-century African American who was a women's rights activist and an abolitionist. It was chosen by Valerie Ambroise, then aged 12, from Bridgeport, Connecticut. She won a worldwide essay contest for young people to name a rover for an accomplished woman.

Pictures sent back by previous spacecraft have shown signs of surface water on a much earlier Mars. Life on Earth developed in the presence of water. So one of the major objectives of 21st-century flights will be to search Mars's now-dry features for signs of life. Finding them would change the way scientists think about life elsewhere in the universe.

Mars Odyssey

The Odyssey spacecraft, an orbiter programmed to map the entire surface of Mars, returned its first high-resolution photos early in 2002, indicating the presence of frozen water on and below the planet's south pole.

MARS EXPLORATION ROVERS

Several missions to Mars are planned for 2003 because Mars and Earth will be very close together. The National Aeronautics and Space Administration (NASA) is scheduled to launch two advanced robotic rovers in June 2003, landing in January 2004. The rovers are identical and their designs are based on Sojourner's.

Each rover will land in a different region where research already shows that life may have existed at one time. These six-wheeled, 300-pound rovers will evaluate the composition and texture of rock and soil samples. They are designed to travel a distance of 330 feet per sol, beginning near the landing sites. The work is expected to continue for at least 90 sols. The results will help build a history of climate and water at the sites.

The rovers' six instruments will examine each sample, many of them improved versions of those carried on Sojourner. They are

- an **abrasion tool** that can scrape a rock to expose fresh surfaces for study,

- a **panoramic camera,** to take visible color and infrared wide-angle pictures, to show the geologic context,

- three German-built **spectrometers,** to determine the composition of soil samples—one to discover the various samples' abundance by determining their heat characteristics, a second that identifies iron-bearing minerals, to show whether the materials were formed by water, and a third that tracks a sample's alpha particle/proton/X-ray signatures, and

- a **microscopic imager,** to show the samples' fine-scale features, such as size and angularity; these are clues to their formation and whether they have been moved to their present location by water.

MARS EXPRESS

Also in 2003, the European Space Agency plans to launch "Mars Express," a British-led effort to land on Mars to search for life. Organic matter found on some meteorites of Martian origin may show traces of life-forms. Mars Express will also perform geochemical and atmospheric analyses. A rover called Athena, which will photograph and perform inorganic analyses on rocks and soil, may also follow it.

This *exobiology* (biology outside of Earth) mission will consist of a spacecraft and a lander. The lander is named Darwin 2, in honor of the British scientist Charles Darwin, whose 19th-century journey on the ship HMS *Beagle* led to his book, *On the Origin of Species.*

Missions to Venus

In 1961, the Soviet Union launched two spacecraft toward Venus, the first in a decades-long series. In 1962, the United States began launching its own series toward the planet. Just six of these ships—the Soviet Union's *Venera 9* and *10* in 1975, *Venera 11* and *12* in 1978, and *Venera 13* and *14* in 1981—successfully orbited Venus and put their landers on its surface.

The landers were hermetically sealed to protect the instrumentation from Venus's hot temperature (over 840°F) and high pressure (eight times the pressure on Earth). Instruments on *Venera 13* and *14* included camera system, X-ray spectrometer, a drill and other surface sampling tools, a penetrometer (to measure the type of surface soil), and a seismometer.

Each lander took soil samples and deposited them in the instrument chamber. It took chemical and isotopic measurements, monitored the spectrum of sunlight, and recorded electrical discharges (lightning). Their results were relayed to the orbiters and then to Earth.

The Venera landers provided the basic information that could lead to more sophisticated exploration of Venus in the future.

Solar System and Beyond— Probes and Observatories

Autonomous spacecraft are also exploring other parts of the solar system. The Sun is being observed by SOHO (the Solar and Heliospheric Observatory) and Genesis. And NEAR Shoemaker has landed on the asteroid Eros.

SOHO (SOLAR AND HELIOSPHERIC OBSERVATORY)

SOHO is a cooperative mission of the European Space Agency and NASA to study the Sun's dynamics during an entire 11-year solar cycle. During some parts of this cycle, the Sun is very active, emitting large amounts of electromagnetic radiation (cosmic rays) and the solar wind (plasma, or ionized particles, constantly emitted from the Sun's surface). At other points of the cycle, the Sun is relatively quiet. SOHO is constantly recording the Sun's activity and beaming the information back to Earth.

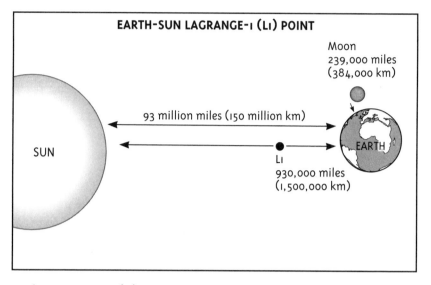

Earth-Sun Lagrange-1 (L1) Point

The 4,000-pound SOHO was designed and built in France and launched by NASA in 1995. It is part of an international program called the Solar Terrestrial Science Program.

SOHO circles the Sun at an altitude of 930,000 miles above the Earth—three times as far as the Moon's orbit. SOHO's position is a point called L1 (for first Lagrange point)—the place where the Earth's gravitational pull is the same as the Sun's. This point shifts slightly due to changes in the orbits of other planets, so SOHO is in a *halo orbit* around L1. SOHO revolves around the Sun with the Earth.

The spacecraft's guidance system stays focused directly toward the Sun by calculating the readings from a sun sensor and a star tracker. Scientists use SOHO's 12 instruments, including cameras and telescopes, much the same way that they would from a ground-based observatory.

NEAR SHOEMAKER

The NEAR (Near Earth Asteroid Rendezvous) mission was the first in a new generation of small and relatively inexpensive NASA spacecraft. It was launched in 1996 to fly past one asteroid and land on another, 433 Eros, photographing the descent. "Near Earth" means that the asteroid's orbit crosses that of the Earth. The name *Shoemaker* was added to honor an eminent astronomer, Gene Shoemaker, after his death.

The spacecraft was designed to operate autonomously, using commands sent from Earth and stored on board. It was in contact with Earth only a few times a week to receive commands and send back the data it had collected. If it sensed any problems, it was designed to go into a "safe" mode and wait for commands to correct them.

Eros is an unusually large baking-potato-shaped asteroid $25 \times 9 \times 9$ miles in size. From Earth-based astronomical studies, it has the same surface composition as many asteroids. The mission was designed to answer several important scientific questions, such as, How do Eros's surface characteristics compare to those of planets? Is it a solid fragment of a larger body or a pile of rubble? Is it related to any known meteorite types? Does it have a magnetic field? Does it have any satellites? What are they like?

The spacecraft weighed 10,000 pounds on Earth and directed the use of six on-board scientific instruments:

- a **magnetometer,** to find out whether Eros has a magnetic field,
- several **spectrometers,** for measuring the elemental and mineral composition of the asteroid's surface,
- a **spectrograph,** for measuring the spectrum of sunlight reflected from Eros's surface in the near-infrared range,
- a **laser rangefinder,** to closely map the asteroid's shape, and
- a **multispectrum imager** to study its geology.

After orbiting and photographing Eros for a year, in February 2001, NEAR Shoemaker became the first spacecraft to land on an asteroid. It made a gentle landing that left the spacecraft still operational. This let the scientific instruments analyze the surface from only four inches away. In all, NEAR Shoemaker took more than 160,000 detailed pictures of Eros.

GENESIS

Despite all the space research performed since the 1960s, several questions about the solar system's origins are still unanswered. For instance, How did the planets originate? Why do some planets, such as Venus, have thick atmospheres? And at what point did the Earth become suitable for life? The key to answering these questions is in the amounts of various elements in the solar nebula—the Sun's early, primitive form. If scientists knew this, they could compare them to the same elements in planetary atmospheres. The variations will tell them much about the solar system's early formation.

Scientists think that the Sun's outer surface is much like the original solar nebula. This outer surface is constantly throwing off particles—what is called the solar wind—which reaches far out into the solar system. They have designed the Genesis spacecraft mission, launched in 2001, to capture samples of this wind and return them to Earth for detailed study.

The robotic Genesis spacecraft is equipped with ion and electron monitors to determine the characteristics of the solar wind. After a three-month flight, the Genesis craft will go into a halo orbit around L1.

Its disk-shaped sample return capsule is about five feet in diameter and four feet high and weighs about 500 pounds. It contains three collectors and a concentrator. This is an electrostatic mirror that concentrates oxygen, nitrogen, and the noble gases by a factor of 20. Over the next 30 months it will capture solar wind samples and seal them into a contamination-proof canister. In 2004, the capsule is scheduled to return toward Earth and open a parachute. About one and one-half miles above the Earth, a helicopter will retrieve it. Then it will be taken to the Johnson Space Center for distribution to research laboratories.

Aerial Robotics

Some flying robots are also being tested for use in two environments—space and Earth's own atmosphere. These "aerial robots" include the Autonomous Helicopter and the Helios Prototype.

THE AUTONOMOUS HELICOPTER

This small low-flying machine was developed at Carnegie Mellon University. It is a 160-pound, 14-foot-long aircraft that is based on a remote-controlled crop duster built in Japan by Yamaha. It navigates by using Global Positioning System Satellite information and a video system that senses dangerous landscape features. Its primary function is to map a land area.

It can also search for people or features by color, as in rescuing someone at sea. It uses a preloaded flight plan or one that is sent after it is airborne.

One advantage for its possible use in space is its low manufacturing cost—about $100,000. The helicopter has been tested on a Mars-like island in the Arctic, and some day it might be part of a Mars mapping mission.

HELIOS PROTOTYPE

This is an ultralight robotic "flying wing" that can go as high as 96,000 feet. Powered by solar panels, it can remain aloft for extended periods of time, either mapping land areas or taking samples of the atmosphere. For instance, it could closely monitor pollution movements and ozone depletion.

Further work by its developer, Aerovironment, Inc., might result in letting it fly as high as 100,000 feet (Earth-atmospheric conditions are similar to those in the Martian atmosphere).

7

ROBOTS, SOCIETY, AND YOU

Since early in the last century, people have envisioned robotic servants around the house, vacuuming, mowing the lawn, serving cool drinks on hot days, and generally making life easier for the human occupants. Up until the 1990s, the few home-based robots were usually made by technically minded people and advanced hobbyists. Now, there's a whole category of service robots for personal and private use. It includes domestic robots, such as lawn mowers, and entertainment/hobby/leisure-time robots.

The second group includes toys and doglike companions and has expanded to include organized robot competitions. Sumo wrestling and other types of battle tournaments, complete with celebrity sponsors and participants, are shown on TV. They're even used in commercials for other products.

There are robots for every age group, from dolls and other toys for small children to trainable petlike "animals" that appeal to all ages. There are also personal robots in one of the hardest designs to reproduce—the human body.

Humanoid Robots

The first "robots" that looked like humans were developed in the mid-20th century. Though they had two arms, hands, legs, and feet, they were more like automatons than robots. Built before the widespread use of the computer, they lacked the intelligence to interact with people. They could follow only limited directions and their movements were stiff and jerky.

Emulation of human movements was a far more difficult problem than robot scientists had believed. Only in the year 2000 did robots with humanlike movements, interactivity, and usefulness start to become available. Development of *humanoid robots* has taken place mainly in Japan. The reason goes beyond general robotic development. In Japanese culture, family members have had a deep traditional obligation to care for their aging relatives. Now the country faces an aging population, with fewer family members close by to care for them. Humanoid robotic personal assistants are expected to take over some of this obligation and also assist in security and disaster rescue operations.

Though similar in appearance, humanoid robots have been designed for different purposes.

ASIMO (ADVANCED STEP IN INNOVATIVE MOBILITY)

Honda Motor Co., Ltd., designed ASIMO as an aid to people with limited mobility, turning their lights on and off, opening and closing doors for them, and pushing their wheelchairs. Future models will be able to bring food to people and generally assist around the house. ASIMO can recognize voice commands and the direction they're coming from and obey them to move its arms, hands, and to walk. ASIMO interacts so easily with the public that Honda now rents the robot for use as a guide at conventions and trade shows.

ASIMO, which is powered by Ni-MH batteries, carries its controller in a backpack and looks very much like an astronaut in a space suit. It stands about four feet tall and weighs about 100 pounds. Its height was chosen so ASIMO can work with office equipment, kitchen counters, and other typical living spaces. Also, people have to look down at it, which makes interacting with it less overwhelming.

ASIMO has humanlike joints for natural movements of its head, arms, hands, legs, and feet. ASIMO's neck has two DOF, allowing its head to move up/down and rotate. Each of its humanlike arms has a total of five DOF. Its three-DOF shoulder joint moves the shoulder forward/backward and up/down, and it rotates. Each elbow has one DOF for moving forward/backward, and each rotating wrist joint has one DOF. ASIMO's hand provides a total of one DOF to its five grasping fingers. Each hand has a grasping force of about one pound.

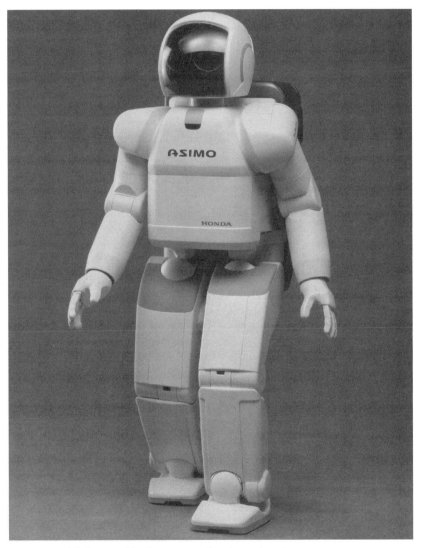

ASIMO, Honda's humanoid robot-companion, in motion [Honda's ASIMO Humanoid Robot—courtesy of American Honda Motor]

Each leg has six DOF. The three-DOF hip joint moves forward/backward and left/right, and it rotates. Its one-DOF knee moves forward/backward. And its two-DOF ankle moves forward/backward and left/right.

ASIMO walks at speeds up to one mile per hour, with variable *stride* (time between steps) for climbing stairs, walking on uneven terrain, such as hills, and turning corners. Its flexible i-WALK technology allows naturalistic autonomous locomotion, without the hesitation between strides found on earlier robots. The robot is also able to shift its entire center of gravity as it turns, for instance, learning into the curve when it turns a corner.

Unlike the stored walking patterns used by earlier robots, ASIMO uses a real-time "predicted movement control" to navigate on new or irregular paths. The robot sets its departure and termination points, along with several intermediate ones. This system keeps the robot from going off course and also corrects its foot placement and body direction one step at a time.

SDR-3X

Sony Corporation developed SDR-3X as an entertainment robot. It can perform basic humanlike motions, including walking, changing direction, getting up, kicking a ball, balancing on one leg, and even dancing. The robot has 24 DOF—two DOF on the head, two DOF on the body, four DOF on each arm, and six DOF on each leg. It is 20 inches tall and weighs 10 pounds. Its central processor uses two 64-bit RISC chips.

The robot recognizes about 20 verbal commands through microphones in both "ears." It responds to these commands with synthesized speech. It can also be operated by remote control from a Wireless LAN card in a personal computer's card slot.

It moves in response to its image recognition system as well. It can tell one color from another and recognize an image, based on data captured by the CCD camera that serves as its eyes. It plays a ball game that takes advantage of these abilities, recognizing a ball of a specific color, finding the goal net, moving beside the ball, shooting the ball, and determining whether or not the ball has landed in the net.

SDR-3X uses the operating system, called OPEN-R, that Sony has developed for all of its autonomous robots. The operating system is modular, giving each robotic function (such as speech recognition) a separate program, which makes it easy to upgrade functions or add new ones. Programs are stored on 16 MB Memory Sticks.

OPEN-R controls SDR-3X's bipedal (two-legged) walking with two technologies—first, the actuator that moves the joints and, second, what Sony calls the Whole Body Coordinated Dynamic Control, which provides real-time control of the joints.

Each actuator ("Actuato") is composed of three units—a motor, a gear, and a circuit—providing high-power output from a lightweight device. Actuators come in three sizes, measured in torque rate, size, and mass, depending on joint size and the power output it needs.

For instance, the knee joints require high-power output and large size, mass, and torque output to support the physical stress of walking. As SDR-3X walks, it moves its body in the same way that humans do, the upper half of the body moving to counteract the *yaw*-axis created by the lower half. SDR-3X can walk at a speed of 0.5 miles per hour, with 2.5-inch strides.

As it moves, it keeps from falling over by evaluating the information it collects from a posture sensor on its torso and touch sensors on the bottoms of its feet. Other sensors include an infrared distance sensor, two *accelerometers* that detect acceleration, two angular rate sensors, and eight contact (touch) sensors.

SDR-3X's movements can also be programmed on a computer using simulation software.

H7

This battery-powered humanoid robot was designed to be used in research in robotic artificial intelligence. It was developed at the University of Tokyo and built by Kawada Industries, Inc. It uses the RT-Linux operating system, which is designed for robotic research such as artificial intelligence, sensing, and systems control.

H7 achieves full body motion with a total of 36 DOF, including six DOF for each leg, one DOF for each foot (a toe joint), seven DOF for each arm, one DOF for each gripper, two DOF for the neck, and three DOF for the eyes. It is five feet four inches tall and weighs about 120 pounds. Its on-board controller features two 750 MHz Pentium III processors, and it can be connected to a network by wireless Ethernet.

Applications under development for H7 include intelligent software for real-time operations, dynamic walking pattern generation, 3-D vision systems, and motion planning.

Robots as Toys

Technically complex robots have become popular as toys (for children) and for home entertainment (for adults). Some are doglike or based on other real or imaginary animals, while others are humanoid. Many of them have advanced artificial intelligence programs that let them grow from babyhood to maturity, develop personality traits, and learn to recognize people they interact with. Also, their behaviors trigger emotional responses in the people they interact with.

The number of behaviors and interactions available on a robot generally increase with the price and sometimes with the age group it targets. One of the most advanced, popular, and costly of these robots is Aibo, a doglike robot from Sony.

AIBO

Aibo (Japanese for companion) is a small four-legged intelligent robot—six inches wide, 11 inches long, 10 inches high, excluding ears and tail—developed by Sony Corporation. Called an entertainment robot, it is designed to interact with its human owners and, through its behavior, solicit an emotional response from them. Though doglike in

Furby robotic toy, with its own language called Furbish, capable of learning English, developing behaviors, and interacting with humans (Image courtesy of Tiger Electronics)

Meow-Chi, a palm-sized robotic cat toy that communicates, senses light, sound, and touch, and can respond, adapt, and change [Image courtesy Tiger Electronics]

appearance, Aibo can be thought of as maturing from baby to child to adult, as well as from puppy to dog. In addition, when it is connected to the owner's computer, it can send, receive, and read e-mail and keep track of appointments. Aibo comes in a choice of three colors, gold, silver, and black.

Even though it costs about $1,500, 120,000 Aibo robots were sold in its first year, 1999, mostly in Japan, where it is used as a companion for older people and for children in hospitals, for example, as well as entertainment. Though smaller numbers have been sold in the United States and other countries, Aibo has become the world's most popular robot.

With its 20 DOF and the proper software, it can follow a ball and play with it, bow down like a dog to solicit interaction, and obey spoken commands like Sit and Lay Down. Its owner can use a system of encouragement (patting its head and saying Good Boy or Good Girl) and scolding (or slapping a newspaper near it) to change its behavior from puppyish to grown dog. The robot will develop and display six different moods, indicated by changes of color on its "eyes" and tail.

Its interactions with people include rubbing its face against the cheek of someone it recognizes. When its owner picks Aibo up and says "Good

Aibo, a doglike robot, ready to interact with its owner [Sony Electronics, Inc., Entertainment Robot America]

night," the robot becomes sleepy and changes to its sleep mode. When scolded, it hangs its head and registers a change in mood. Depending on its mood, Aibo might not respond to a question or command.

It can also answer questions by responding with its eye and tail-lights. For instance, in response to the question How old are you? it will flash the correct number of times. It can imitate what a person says to it as well.

Aibo's 20 DOF include one DOF for the mouth, three DOF for the head, three DOF for each leg, one DOF for each ear, and two DOF for the tail. It weighs about 3.3 pounds. It runs autonomously for about 1.5–2.0 hours and can use regular batteries, a lithium ion battery pack, or an AC power adapter. To interact with its environment, it has built-in sensors for temperature, distance (infrared), acceleration, pressure (sensors on its head, chin, and legs), and vibration.

Aibo "sees" with a *Complementary Metal Oxide Semiconductor (CMOS)* image sensor and, on command, can also take a picture of what it views. Its head has a microphone to receive voice commands and a speaker to respond.

The robot uses Sony's OPEN-R operating system, and its programs are stored on a memory stick that is inserted in a slot on Aibo's stomach. There is also a Type 2 PC card slot near the tail for uploading and downloading data.

MY REAL BABY

My Real Baby, an animatronic robot developed by Hasbro, is designed to give girls age three and up a dynamic baby doll that realistically interacts with them, develops, and responds to the child's interactive nurturing. Real learning does not take place, though.

The animatronics gives the robot a human-baby's appearance and comes with either light skin (made of rubber) and light wavy hair or with dark skin and dark curly hair. The robot is 18 inches tall, 10 inches wide, and four inches from front to back, and it weighs about 3.1 pounds. Its power source is six AA batteries.

My Real Baby had its origins in the iRobot Corporation, founded by behavioral robot scientist Rodney Brooks. He also developed the Behavior Language Operating System, which lets the robot respond naturally to the child's actions—if the child doesn't feed it, it cries, for instance. The robot uses the operating system's Version 6.0.

The operating system works with what the manufacturer describes as Natural Reaction Technology, a proprietary artificial intelligence system that simulates the natural expression of hundreds of behaviors, often in parallel. It has both awake and asleep modes, billions of four-sound utterances, spoken words, and phrases, and dozens of sensors throughout the doll's body to enhance realistic responses. It can be reset to the earliest stage of development.

The robot can smile or frown, giggle when its feet are tickled, and laugh when it is bounced on the child's knee. Speech begins with single words, like *Mama* and *yum-yum*. This is followed by short sentences, such as "I love you, Mama." Rocking puts the robot into its sleep mode. (There is also a button that immediately places the doll in the sleep mode until the button is pressed again.)

As the robot develops, its behavior expands into unpredictability and role-playing and the robot initiates its own actions, behaviors, and preferences.

For enthusiasts who don't want to purchase a fully assembled robot, there are kits that contain all the parts required for a working robot, along with instructions for assembling it.

Robot Kits

A robot kit—available from robot hobby stores or toy stores—gives the builder the pleasure of quickly having the robot ready to operate. Assembling a kit helps a newcomer understand the basics of electronic circuitry and the robot's working parts. The simplest kit robots perform basic movements: starting, stopping, moving backward and forward, or following a wall. Many beginner and advanced kits use a variety of modern technologies with regular upgrades.

BEGINNER KITS

Even beginner robot kits contain sensors, legs, wheels or treads, and interchangeable parts that allow different configurations. One very simple robot moves forward until it bumps into something, then reverses. Other kits let the enthusiast build simple robots that can still compete with each other in informal contests. Some kit robots use motors and battery power. Others have infrared detecting beams and moving arms for shoving an opponent.

ADVANCED KITS

More complex robots have sensors, allowing them to react to their environment. Sound sensors register voices or hand claps. Other sound sensors detect *ultrasonics*, noises too high for human ears to pick up. These features let inexpensive toy robots perform the same tasks that commercial robots do. Infrared sensors, for instance, let some robots follow dark lines on a light surface—the same principle used by some robotic service carts that deliver mail or supplies in commercial buildings. Touch sensors stop a robot's motion when it bumps into something or allow the robot to detect obstacles and detour around them.

The most advanced kit robots can be programmed. Some use a keypad for entering commands. Some come with software that can be downloaded from a personal computer. In some cases, the user can also write command software. One omnidirectional robot, for instance, programs fuzzy logic instructions, allows if-then decision making, and can be connected to a personal computer for extended programming. Another connects to a computer for programming, can be controlled by a mouse, and uses its video system to map and name the areas of the house that it navigates in. It also has enough pulling force to haul a wagon filled with snacks and drinks.

Lego Mindstorms produces one of the most complex and flexible kits.

LEGO MINDSTORMS KITS

In 1980, Seymour Papert, then a professor of mathematics and education at MIT, wrote a book called *Mindstorms* about children and computers. He developed an easy-to-use computer language called Logo, which he adapted from the AI language LISP. Then he applied it to a robotic, pen-carrying turtle that children could easily program—with commands like MOVE, TURN, and FORWARD—to draw complex graphics on paper.

Lego is a Danish company famous for the simple interlocking building blocks that children and adults use to create buildings, both simple and elaborate. It also partners with the MIT Media Lab to create robot kits, named Mindstorms in honor of Papert's book. The Lego Mindstorms Invention System consists of Lego blocks (including bolts, wheels, axes, and wires), a programmable computer "brick," touch and light sensors, motors, and gears. It is popular with people of all ages.

The computer brick, which stores five programs at a time, uses an infrared transmitter to download programs. It can also communicate with other brick computers. Instructional material shows the user how to build and program a robot. The Mindstorms programming language builds its programs in a series of modules, paralleling the process of constructing a structure as a series of individual blocks. A program can be constructed on a computer, where each command is visualized as a block added to the preceding command. Programs are easily improved by making changes within a module or adding new modules.

The basic system is flexible enough for the new enthusiast to easily build, for instance, a robot with a robotic hand that opens and closes.

The Mindstorms Visual Command is a digital camera (capable of capturing video and still images) and microphone attachment for the basic system. With it, the robot can respond to motion, color, or light. Other add-on systems include the Droid Developer Kit and the more advanced Dark Side Developer Kit, both of which let the builder construct *Star Wars*–related robots.

Building Robots from Scratch

If you want to construct a robot entirely on your own, here are some hints to help you get started. A hobby-robot magazine such as *Robot Science and Technology* can be very helpful.

Most people begin their first robot by deciding what they want it to do. After defining the task, the builder should make a list of the parts needed.

For instance, a mobile robot will require wheels or treads. For a walking robot, you'll need to decide which design and method of moving the legs to use. If the robot will grasp, lift, or carry something, it must have an arm and hand (manipulator and end effector).

You must also decide what the robot's world will be: where it will operate and what, if anything, it must know about its environment. If it should avoid objects, you'll need a touch sensor and perhaps an infrared sensor.

A knowledge of electronics is important. Large bookstores, technical bookstores, and some electronics stores carry books on the basics of electronics. Electronics stores are listed in the Yellow Pages under *Electronics Equipment and Supplies—Dealers*.

The builder also needs a way to control the robot. One way is with a joystick, like those on some computer game consoles. For a larger number of commands and more flexibility, you may want to put your commands in a computer program and download it to your robot. If you use a transmitter, you can send wireless commands to your robot.

Many people write robot programs in BASIC. Others write programs in Logo or in the AI language LISP.

Robot Contests

How does your home-built robot compare with similar ones? A robot competition is one way to find out. Nationwide or international contests for soccer-playing robots, sumo wrestling robots, firefighting robots, and others take place every year. For contest rules, see the websites listed in the Appendix.

ROBOCUP

RoboCup is an annual worldwide competition for soccer-playing robots. In 2001, it took place in Seattle, with 111 college teams representing 23 countries. Unlike the human game, in robot soccer the ball never leaves the ground. In 2002, teams of humanoid robots will compete in "humanoid league" soccer matches in Japan at the same time as the World Cup finals.

Hiroaki Kitano, a researcher at Sony Computer Science Laboratories, who chose the soccer format because the game is played virtually everywhere in the world, founded the event in 1996. Its purposes are to create the best robots and programming for winning at soccer and to promote teamwork between the human owners.

RoboCup robots competing in Seattle, in 2001 [© 2001 The RoboCup Federation]

After each year's contest—consisting of a round-robin contest, followed by a play-off—is concluded, the teams are required to make their programming public.

Four classes of robots take part—small (seven inches), middle (20 inches), Sony Aibo robots, and computer-simulated. RoboCup's long-term goal is, by 2050, to assemble a team of autonomous robots that can defeat the human team that wins the World Cup.

BATTLEBOTS

BattleBots are remote-controlled robots whose battles are shown on pay-per-view TV. The audience includes fans of wrestling, boxing, and combat sports, and computer and technical enthusiasts. Robots compete in one of two weight classes, either Stomp-Bot Heavyweights (174–315 pounds) or Super Heavyweights (up to several hundred pounds). They first compete one-on-one in three-minute contests, then in a free-for-all among all robots in the same weight class. Anyone is eligible to enter a robot. BattleBots also operates a competition called BattleBots IQ for high school students.

BattleBots-Biohazard vs Overkill in a heavyweight bout during the BattleBots May 2001 tournament [Photo courtesy of BattleBots Inc./© 2001 Battlebots Inc./Daniel Longmire, photographer]

A sumo robot, ready for a match [David E. Calkins]

Robots can use tactical tools, such as hammers, axes, spokes, and buzz saws. Also, they compete in a ring filled with pop-up hazards such as sledgehammers, which try to destroy the robots before their opponents do.

SUMO WRESTLING

Sumo is a traditional Japanese sport in which opponents try to push each other out of the wrestling circle. Sumo wrestling robot competitions are held in many cities in the United States, Canada, and Japan.

As in the human version, a robot match of three minutes starts with the competitors bowing to each other. In the contest, they use robotic strategies such as camouflage, disablement (shining a flashlight in the opponent's vision system, for instance), and boundary detection to defeat the opponent.

Matches take place in a five-foot-diameter ring. They typically have two classes, Japanese Sumo Class, with robots about eight inches square and six and one-half pounds, and Mini Robot Sumo Class, about four inches square and one pound. Each class also has two divisions, autonomous and remote control.

FIREFIGHTING ROBOT CONTEST

Each year Trinity College, in Hartford, Connecticut, sponsors a fire-fighting contest for home robots. The contest's goal is to build the robot that can move through a miniature house, find a lit candle (representing a fire), and extinguish it (with water, air, carbon dioxide, or other safe substance) in the shortest time. The larger goal is to advance robotics technology for the participants and the world in general.

Robots should be no bigger than 12.5 × 12.5 × 12.5 inches, but there are no weight restrictions. The search for the candle should take no more than five minutes. As a way of encouraging navigation methods other than dead reckoning, robots that use other methods will gain extra points. The six winners in each division—junior (grade eight and below), high school (grades 9 to 13), and senior (college students and adults), expert, and commercial—receive cash awards.

FIRST CONTESTS

The nonprofit organization FIRST (For Inspiration and Recognition of Science and Technology) administers two contests for young people, one for high school students and one for ages nine to 14. Founded in 1989, FIRST had a 2002 schedule that included more than 20,000 students in 650 teams. The competitions emphasize goal-oriented robotic activity and team cooperation.

Competition for High School Students. A team of students and their adult mentors must design and build a robot from a standard set of parts (provided by FIRST) during a stated six-week period. The average team has 35 students and six mentors. FIRST says winning involves excellence in design, team spirit, professionalism, and the ability to overcome obstacles. In 2002, teams competed in 17 regional competitions in the United States, Canada, Brazil, and the United Kingdom. More than $1.6 million in college scholarships are available to participants.

Lego League Competition for Nine- to 14-Year-Olds. The contest for teams of 10 students combines sportslike competition with academic ability. Each team designs, builds, programs, and tests its robot, then enters state competitions.

The next chapter will discuss the future design and use of robotics and some of the ethical questions surrounding it.

THE FUTURE
OF ROBOTICS

Robots are so important a part of our present that their roles will undoubtedly be even larger in the future, in cyberspace as well as in the real world. They will be making things in factories and selling them to us on the Internet. They will be our coworkers on Earth and the explorers of our neighbor planets. Their traditional role has been in manufacturing. How will robotics affect our jobs? What will they mean to the U.S. economy? And a final question: Should there be any limit to what robots can do?

Robotics Industry

The modern robotics industry began in the United States in the 1980s, but the United States has been slower to use manufacturing robots than other countries. Where is robotics headed? An answer comes from the *2001 World Robotics Survey*, compiled by the United Nations Economic Commission for Europe, in cooperation with the International Federation of Robotics, and based on information from national robot associations. The survey projects steady worldwide growth in both manufacturing and personal robotics, based on present economic patterns. Both categories increased in use during the 1990s, a trend that is projected to continue.

ROBOTS IN MANUFACTURING

In 2000, 750,000 robots were working in manufacturing. Half of them were in Japan, 27 percent in Europe, and 12 percent in North America. By 2004, the Economic Commission's survey expects an increase to 975,000 industrial robots worldwide—with 46 percent in Japan, 32 percent in Europe, and 12 percent in North America.

In Japan, in 2000, there were almost 300 robots per 10,000 workers, the highest rate in the world. (Japan includes all types of robots, while all other countries count only multipurpose industrial robots.) In Europe, there were between 50 and 120 robots per 10,000 workers. In North America, the rate was 49 robots per 10,000 employees. From

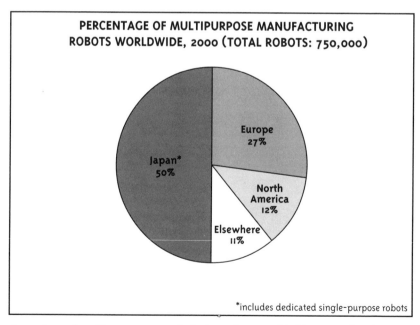

Percentage of multipurpose manufacturing robots worldwide, 2000 (total robots: 750,000) [Source: *2001 World Robotics Survey*, UN Economic Commission for Europe, 2001]

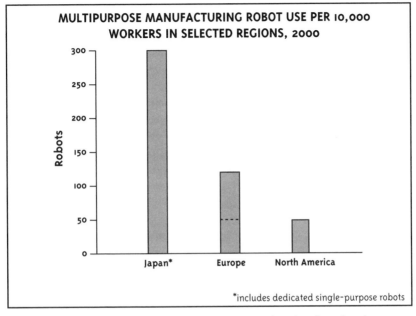

Multipurpose manufacturing robot use per 10,000 workers in selected regions, 2000 [Source: *2001 World Robotics Survey,* UN Economic Commission for Europe, 2001]

2001 through 2004, sales of robots were expected to increase 7 percent per year, from 99,000 in 2000 to 128,000 in 2004. Excluding Japan, the annual increase would be 8 percent.

The highest concentration of robots is in the auto industry. In Italy and Germany, there is almost one robot for each ten production workers.

Robotics has shown the same pattern of pricing as the rest of the computer industry. Prices in 2000 were less than one-fifth of those in 1990 for robots of comparable performance. Robots paid back their original costs in one or two years. In the United States, when improved quality is included, robots in 2000 cost just 12 percent of their cost in 1990.

Part of the favorable economics is due to rising labor costs—as workers' wages rise, robots become more cost effective. Also, robotic technology has continued to improve. For these reasons, robots are increasingly used in both developed nations and in developing countries, such as Brazil, China, and Mexico.

SERVICE ROBOTS

This classification includes both professional robots, such as underwater ROVs, surgical assistants, and laboratory robots, and personal robots, including toys, lawn mower robots, and hobby robots. Underwater robots account for more than 60 percent of service robots' total value, mainly because each one is so expensive.

In 2000, there were more than 10,000 service robots worldwide, worth more than $1.715 billion. By 2004, the commission members believe that there will be more than 30,000 service robots, worth $2.4 billion.

Robotic improvements in the near future will be evolutionary, with an increase in the kinds and abilities of robots already in use, both commercially and in the home. In the more distant future, experts expect new and revolutionary designs of robots, such as those designed as artificial life and swarms of nanoscale and microscale robots, to come into use.

Robots in the Near Future

In the future, robots will be smarter and make more use of "telepresense" than current robots do. NASA, for instance, plans evolutionary improvements in its planetary orbiters and landers.

Landers may be designed to cover comparatively long distances—a dozen or more miles in a year's time—as it uses an extensive pack of scientific instrumentation to analyze the Martian surface.

A more versatile version of the same rover might also be used to collect rock and soil samples. It would then launch the sample package into orbit with a small rocket, to be retrieved by a planned U.S.-French orbiter. Finally, the orbiter would return to Earth. The French space agency is planning to test its sample orbiter in 2007.

Also in 2007, NASA may launch another small, inexpensive Mars lander that takes photos on its way to the surface. Then it might use other instruments to analyze its landing site. Such a lander may also serve as a test for future missions to the two Martian moons, Phobos and Deimos.

One of the most advanced robots being tested in the early 21st century is Hyperion, developed at Carnegie Mellon University (CMU). This Earth-based robot is a model for possible work at high latitudes on Mars and other solar-system bodies. It uses artificial intelligence to be aware of its environment, plan and carry out navigation and scientific tasks, avoid obstacles, and track the Sun to give itself a constant power supply over many months or even years.

The robot has four motorized DOF, one for each of its four independently controlled wheels, and one passive (nonmotorized) DOF that rolls and yaws. Independent control of each wheel keeps the robot from skidding, which wastes energy. The robot is six feet long, six feet wide, and eight feet high and weighs about 350 pounds.

In early Greece, Hyperion (meaning "he that walks on high") was the sun god. The robot Hyperion uses 35 square feet of solar panels to generate energy and batteries to store any excess power. It synchronizes its movements with the Sun's path to maximize solar power generation. In uncertain situations, such as unexpected physical obstacles, it can determine whether it has enough power to avoid or overcome them. For instance, if it confronts an unexpectedly rocky terrain, it can

Hyperion Mars robot, shown on Earth [Carnegie Mellon University]

determine whether it has enough power to plot an alternate path and follow it.

Hyperion also can regulate the temperature of its components. To cool them off, it can fall back behind the night-to-day terminator (the line that separates night from day), after making sure that it has adequate power. To maintain or increase temperature it can move with the terminator, so that it is partly in the shade and partly in sunlight or ahead of it.

To see the terrain, the robot uses a wide-angle camera and two standard-angle digital cameras. It uses computer algorithms to evaluate what it sees, such as obstacles, and draw up alternate routes. It also uses a laser scanner as a bumper or fail-safe device, sweeping the path ahead to detect rocks, cliffs, and other obstacles. If the scanner detects one, it signals the robot to stop.

Other sensors include temperature monitors throughout the robot and roll and pitch detectors and odometers to monitor its motions. It also senses its speed and the voltage and current in its motors.

To navigate, the fully equipped robot needs only a terrain map, its estimated location, and the current time.

Hyperion can be either teleoperated or autonomous, using what CMU calls a "sliding autonomy" system. This lets the robot switch from operator controlled to shared operator-robot decision making to full autonomy. For instance, if its sensors tell it that there is a problem, and after evaluation it thinks it can't handle it on its own, it can stop and signal for help.

Hyperion cost only about $100,000 to build, mainly because it uses many commercial components that are certified to work only on Earth. But it has been tested in the Mars-like high Arctic, on Devon Island, in Nunavut Territory, Canada.

Robots in the More Distant Future

Improvements in telepresense—especially vision and touch—will greatly increase the ability of humans to understand conditions at distant locations. For instance, Earth-based humans will be able to understand conditions on other planets. Telepresence will also be valuable when humans land on those planets. Bundled in their spacesuits, they won't be able to directly feel the wind on their faces or touch the rocks with their fingertips. But telepresence, from robotic sensors on their suits, could give them the same experience.

The next step in telepresence could come from what is called "sensory substitution." This is when the brain substitutes one sense for another to provide information. It is well known that people who have vision impairments can use their sense of touch to gain much "visual" information. Braille works on this principle.

Scientists have found that visual information can also be translated into touch information. This would allow cameras on Mars to transmit "touch" pictures through telepresence systems. This could work for telerobots and their operations and also for astronauts on the surface. Sound information can also be translated into touch information. This is now used for people with hearing impairments.

Scientists also see an environmental sensing role for groups of the simple behavioral robots.

Some scientists believe that gesture-based communication will help human operators stay more closely connected to their robots. For instance, arm movements could manipulate a distant robot, rather than a joystick on a console.

An even more direct human-robot connection may be by brainwaves. Scientists have discovered that thought alone can signal a robotic arm. Research with monkeys has shown that as they prepare to take action, such as reaching for some food, their brains send signals that can be detected by electrodes implanted in their brains. Once this information is translated into a computer program, it can be taught to a robot, which can then take action at the same time as the monkey.

Paralyzed humans have also tested such brainpower. They were able to use their brainwaves alone to move a computer cursor.

It is also possible that future robots will be our competitors, rather than our helpers, as fellow living creatures—what's called artificial life.

Living Robots?

Is there such a thing as artificial life—meaning life that exists in a cyberspace that's as real as oxygen-based organisms that walk on Earth? Some people say that there is.

Living species—as defined by biologists—mate or otherwise reproduce, pass on their characteristics to their children, and evolve. Suppose there are other beings that do the same things but that are electronic, rather than chemical, as we are. They exist only in a computer's circuits. Are they alive, the way people, dogs, bacteria, and viruses are?

Until recently, the idea of living robots was just that—an idea. Despite their inventors' efforts, 18th-century automatons were not alive. Since 1950, the computer has made people ask whether machines can think, and some people have answered yes. Now, in the 21st century, will people also answer yes to the question "Can machines be alive?"

Artificial life is a type of computer program that behaves like a living organism, mating (interacting with other programs), reproducing (creating new programs), passing on characteristics according to the laws of genetics, competing with other programs, and evolving into very different programs.

Some artificial life programs use computer graphics to picture their "creatures." Each new generation appears on the screen, showing new characteristics and mutations. Some of them even develop into different species. They become predators and prey. Some species become dominant; others die out.

Some scientists say these "organisms" are very close to being alive. These scientists think (or hope) the programs are erasing the line between living organisms and nonliving matter. If so—and it is a big *if*—the same reasoning could be applied to advanced robots.

One group of artificial lifers thinks that this definition doesn't go far enough, because evolution in the physical world also requires the survival of the fittest—species that can succeed within their environment. They think that their cyberworlds, too, must begin with a network of random nonliving building blocks. In time they reproduce, compete, and evolve. One English researcher named Steve Grand expressed this belief through his popular Internet game involving creatures called Norns. He considers them to be alive.

The main difference, of course, is that the artificial organisms "live" only within the computer—they are electronic. Robots, on the other hand, have always been made of matter and function in the physical world. Now robots are also simulated in the computer. Is the matter-based robot more real than the one on the screen? If a computer-based robot is a form of artificial life, is its matter-based twin also alive?

By even considering such questions, the definition of "real world," some scientists say, is changing. They believe that cyberspace, or artificial reality, is as real as the physical world. If the electronic world is "real," then robot controller brains could contain the computer programs of artificial life. This may seem far-fetched.

But in 2000, scientists at Brandeis University had a robotic computer program create a second generation in the physical world. They gave a list of parts to a robotic manufacturing program. It designed a product, tested and changed it, changed its own programming to reflect this, and then manufactured the final product—a simple wooden device that crawled. No human action was involved.

One lifelike model that scientists are considering for both cyber- and real-world robots is the ant colony. This model is sometimes called "swarm robots."

Ant Colonies and Swarm Robots

Ants in an ant colony have been described as elements of a decentralized brain. Individually, their abilities are primitive, and an individual ant doesn't last very long. As a group, ants are a fine model of efficient foraging. They are able to find food, using the shortest route to bring it back to the nest, and mark and defend the colony's territory. They use powerful pheromones (chemical signals) to summon more ant power. When a few ants find a piece of cake left out on the kitchen table overnight, for instance, it takes them only a short time to pass signals back to the colony. Soon a long stream of ants is heading toward the food. If you look closely, you can see the incoming ants stopping to exchange signals with those that are leaving. Given enough time, they will take the entire food source.

They may not have human intelligence, but in large numbers they often work as if they do, using what's called "swarm intelligence." Computer scientists are using this model of intelligence to improve information transfer on data networks, such as the Internet. One model uses a team of "ants" to move randomly through a database, find specific kinds of information (instead of food), and take it back to the "colony's" home. The result is clusters of related information.

Two scientists at the RAND Corporation's National Defense Research Institute, John Arquilla and David Ronfelt, have suggested military use of swarm intelligence. They recommend creating swarms of soldiers organized into pods and clusters (groups of pods). These rely on continuous networked information as they infiltrate an area, then call in appropriate weaponry without having to go through a traditional chain of command.

Space scientists also are interested in swarm intelligence. Conventional satellites, for instance, are extremely expensive. Using swarm intelligence, scientists could greatly cut the cost by launching a colony of tiny satellites. These would cluster in space and serve the same functions as a single large satellite. Such virtual equipment is expected to be useful both in peaceful space exploration and in military operations.

The next step would be swarm robots (or gnat robots) that would add micro- and nanotechnology to swarm intelligence. There are no working robot models on these scales yet. However, in 2001, scientists at the Weizmann Institute of Science, in Israel, announced the creation of a nanotechnology-scale, programmable computer that uses DNA for hardware, software, and data, and for output that can answer certain yes/no questions. A trillion of these computers fit into a test tube.

In the 1990s, students at MIT's Artificial Intelligence Lab built larger—2.35 cubic inch, 1.8 ounce—ant robots that exhibited a primitive swarm behavior. Each robot was equipped with 17 sensors, including light sensors, infrared receiver, bump sensors, food sensors, and a tilt sensor. The food sensors actually were voltage sensors, because the "food" was crumpled brass foil, which had no voltage. Each robot moved on two treads and had mandibles that let it grasp and lift small bits of food. Each robot also had an eight-bit microprocessor for its brain.

Individual software programs created a series of behaviors, following the subsumption architecture model. For instance, the simple behavior "move to light" consisted of three commands (1) move-forward, the most basic, (2) move-to-light, which overrode move-forward, and (3) move-from-bumps, which overrode the other two. Behaviors, in turn, were grouped into moods.

For communicating, each used infrared emitters, one on the front and one on top. As an example, antlike behavior resulting in clustering was built into the "I found food" behavior. When one ant found food, it sent an "I found food" signal. An ant receiving it moved toward the first robot and transmitted "I see an ant with food." An ant receiving this message moved toward the transmitting ant and also sent the message "I see an ant that sees an ant with food." And so on.

Swarm technology on a nanoscale is still in the future, but robots that emulate an insect and other forms of life have already been tested. They are part of a new robotic field called biomimetics, in which behavioral robots mimic natural animal designs.

Biomimetic Robots

Scientists in many research laboratories are developing small behavioral robots modeled on life forms, often for military purposes. These include Robo-Fly, at the University of California, Berkeley; Robo-Scorpion and Robo-Lobster, at Northeastern University; and also Robo-Lamprey, Robo-Tuna, and others, all designed for specific environments. Once released, they will perform military and hazardous environment tasks such reconnaissance and surveillance, mine detection, fire control, and reactor inspections.

These differ from animatronic robots by their more sophisticated mechanisms and their connection to the habitats of the animals they mimic.

ROBO-FLY

Robo-Fly is a tiny robotic fly, designed and built by a research laboratory at the University of California, Berkeley. Modeled on a fruit fly, Robo-Fly has flapping wings one-half inch across and weighs about

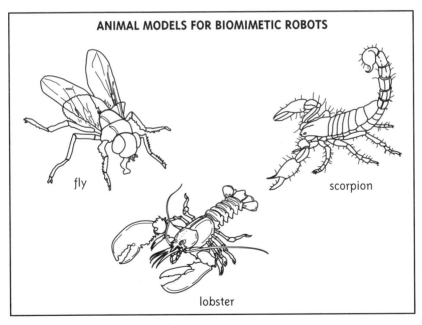

Animal models for biomimetic robots

one-hundreth of an ounce (as much as a flower petal). The wings, which are made of polymer, flap in response to the expansion and contraction of a piezoelectric ceramic when an electrical current is applied. This in turn moves the wings hundreds of times per second. A one-wing model of Robo-Fly was tethered to a boom for its first flight—of about five inches. Free flight tests of complete models are planned for 2003.

The Department of Defense envisions using swarms of Robo-Flies under battle conditions, for measuring the weather or inspecting air ducts.

ROBO-SCORPION

In a project funded by the Office of Naval Research, Robo-Scorpion is under development at Northeastern University. The robot looks something like a scorpion, and its behavior is based on neurobiological studies on invertebrates. Robo-Scorpion is designed to be autonomous, using solar energy for power. Because of its physical appearance, it could take part in stealth operations.

In 2002, Robo-Scorpion is scheduled to be tested in the Mojave Desert. It will make an outbound 25-mile trip and then navigate back to its home base—a total of 50 miles.

ROBO-LOBSTER

A robotic lobster that walks on the ocean floor was developed at Northeastern University's Marine Science Center. Funded by DARPA and the Office of Naval Research, Robo-Lobster is being designed to replace or supplement humans in a hazardous environment such as detecting explosives.

The 18-inch-long robot is battery powered. To walk on its four jointed legs, it uses nickel and titanium wires activated by small electrical charges, instead of muscles. It also has two eight-inch-long oval paddles instead of claws and a tail to make it more maneuverable. It detects obstacles with hair sensors, whose signals are evaluated by a microprocessor "brain." Massa Products Corp., Hingham, Massachusetts, has built prototypes.

ROBOTIC BOWERBIRD

Mimetics is also used in biological research. For example, the satin bowerbird is an Australian species whose male courts a female by building an elaborate, houselike structure (the bower) for her to enter, then mate. The females fly around, examining the males' bowers for characteristics such as symmetry and colorful objects. When a female lands, the male fluffs up his feathers and begins a dance to woo her. In the process, the male must be aggressive enough to keep the female interested, but not so aggressive that he frightens her away.

The University of Maryland's Intelligent Controls Laboratory designed and built several robotic female bowerbirds for the school's Bowerbird Project. Gail Patricelli, a graduate student, used them in the satin bowerbird's habitat to study male-female communication during sexual selection.

The robots were life-sized, with a metal skeleton and realistic feathering. Both their appearance and their behavior were based on videos shot in the field. The satin bowerbird is only about 10 inches long, so the on-board computer (called the "bird brain"), located in the robot's head, consists of a simple eight-bit RISC chip with six input-output lines, 1 kilobyte of program memory, 41 bytes of data memory, and lines for the muscle actuators. Power came from a car battery.

The remote controller, which transmitted from as far as 50 feet from the bower, has separate dials and a switch for initiating smooth head, body, and wing movements. These include lateral left-right motion of the head and neck, a crouch as the body tilts forward just above the leg joint (the bird's center of gravity), and extension of the wings.

Patricelli hypothesized that in real life, the female sends signals indicating her level of comfort with the male's display. The successful males pick up on these and adjust their level of intensity accordingly. By varying the signals in the robotic females that she placed in actual bowers, she proved that this was correct. The female robots themselves were successful, with males attempting to mate with them.

An operating robot always carries an element of uncertainty, perhaps because of imperfect design or construction or of materials that change with constant use or temperature variations. Working with an industrial robot brings special risks, because industrial robots are usually very heavy.

Partnering with a Robot

Working well with a partner requires cooperation. Each person must know what the other person is doing. They must keep out of each other's way. Even when people are "thinking safety," accidents can happen. Lifting and carrying a large heavy object, for example, requires care that one of the carriers isn't bumped from behind by the object or pressed between the object and a wall.

When the partner is a robot, safety is very important, especially in a factory. Robots can be heavy and speedy. Instead of a mild bump, a manufacturing robot can deliver a blow leading to serious injury or even death.

Why do human-robot accidents happen? Sometimes an accident is the robot's fault. It may have moved unexpectedly, the same way a car lurches forward sometimes. Or the robot may drop a tool that hits someone. The robot can make the wrong move because of a computer error or a power failure. Or the robot might be turned on at the wrong time.

People sometimes get careless. They forget that robots are machines. Some of the accidents are the human workmate's fault, like entering the robot's work space. For instance, if a person sees something jammed in a robot's hand, the instinct is to reach over and pull it out. As a result, the robot's hand grasps the human one. Or someone might enter the work space by mistake or for no particular reason. In some cases, workers have been hit from behind and pinned between the robot and another machine or a wall. If the robot is still, the worker may not even realize that it's turned on. The worker causes more than one out of every three accidents. This means that workers need safety training.

But the design of the robot and its work space is also important. Experts recommend operating speed limits. Sensors can be installed to detect humans or objects that shouldn't be in the area. The robot's work space should be fenced in. And there should be switches that stop the robot if someone or something enters the space. Of course, a robot that's designed to operate accurately and reliably is a safety plus.

Designing a robot to be accurate and reliable requires professional skill. Many robotics experts think social responsibility also plays a part: remember the Three Laws of Robotics?

Human-Robot Convergence and the "Three Laws of Robotics" Revisited

Isaac Asimov's Three Laws of Robotics say that robots can protect themselves as long as they don't disobey or harm people. Of course, robots are designed and built by people, so it's up to us to make sure that the Three Laws are designed into robots.

People must also decide how robots will be used. Computer science is already deeply involved in this problem. People have used computers to snoop into people's private files, including those that might be used to program computers. Huge databases collect so much information on each individual that the idea of privacy has all but disappeared.

Computer hackers break into computer files out of curiosity. Others enter files to steal money or information or to destroy or change the information. Such actions are illegal. Each computer designer, manufacturer, and user must make choices. Each person must ask the ethical question, How do I use this equipment to get what I need, without harming someone else or society as a whole?

As robotics plays larger roles in the workplace and at home, the same questions will be asked about it—and some new questions too. When robots can be sent on tasks far from the human operator, who is in control? What will we program them to do? As one scientist has asked, will we just end up as people armed with robots instead of guns?

The Changing Roles of Robots and People

The increasing use of even ordinary commercial robots in the workplace has both good and bad points that people must consider. Robots can do the same job over and over without making mistakes. They work 24 hours a day. Many environments where robots should replace people are obvious—for instance, working in radioactive environments. But according to government figures, even the fast-food industry can be a dangerous place to work.

Every time jobs are taken over by robots, people are put out of work. Robots have already replaced some factory and service workers. The economics of manufacturing says that this trend will continue. Robots can be expensive to buy, but they don't get sick, quit their jobs, need vacations, or want health insurance. For repetitive tasks and automated factories, robots can be a cost-effective choice.

This is especially true in today's global marketplace, where factories in the United States, for instance, must compete with factories in the developing world, where salaries are much lower. Also, some business executives think that today's workers have fewer skills than workers in the past did. The executives say this forces them to use automation and robots.

What happens to the people who are replaced? In the 1920s, the authors of *Metropolis* and *R.U.R.* worried about robots putting people out of work. Today the worries are still with us.

Sometimes the robots are used in the same factory where only people formerly worked. In these cases, replacement is gradual and workers have an opportunity to seek different jobs. In other cases, entirely new factories are built, perhaps in other parts of town, in another state, or in another country. If the displaced workers live in big cities, they may find other jobs in the same industry, but this isn't always possible.

Of course, people can be retrained for different kinds of work. But society must decide who pays for their retraining and education. In addition, if other companies in the area don't need workers, must people move from their neighborhoods, hometowns, or regions to find work?

In the 1960s, American labor unions opposed using robots in factories because they didn't want their members to lose jobs. Now unions often take part in the automation process, helping retrain the workers for other jobs, including installing robots.

As robots take on increased human tasks, look more like us, and think and behave like us, or seem to, will we look at them as workmates? Or will we come to resent them as "people" who work for free and take our jobs away from us.

Robots are changing our lives in other ways too. Psychologists are already worrying about the lower level of creativity in children who play with programmed entertainment robots, compared to those who play with low-tech dolls or action figures and make up their own stories.

It has always been hard to turn away from automation. In the 1740s, the king of France put Vaucanson, who built the flute player and the duck, in charge of making the silk industry in Lyons more efficient. Vaucanson tried to turn the silk weavers' craft into a factory system. But the weavers didn't want to be factory workers. In 1746, they rioted. They even made up nasty songs about Vaucanson and his automated duck. But 60 years later, Jacquard invented his loom in Lyons. His invention made modern weaving factories possible, as well as being a giant step toward computers and robots.

This dance between robots and people, innovation and tradition, is likely to continue. People who want to take part in the world of robotics can learn many things about robots. The experience can also give them the opportunity to learn more about themselves and to have fun doing it.

CHRONOLOGY

4,000 years ago	The first automatons—marionettes used in Egyptian funerals
4,000 years ago	The first manipulators—blacksmiths' tongs
2,500 years ago	The first known robot-related story—Greek myth of Pygmalion, in which a statue is brought to life by the goddess Aphrodite
9th century C.E.	The first mechanized (self-operating) clocks, in the Middle East and Europe
1645	Word "automaton" comes into use
ca. 1723	Pantograph invented
ca. 1738	Jacques de Vaucanson develops his realistic automatons, the flute player and the duck
1801	Joseph-Marie Jacquard invents the punched-card controlled loom
1818	Mary Godwin Shelley writes the novel *Frankenstein*
1830s	Charles Babbage develops his Analytical Engine, a computer precursor
1890	Herman Hollerith's punch-card tabulating machine, the first automatic data processing system, used in the U.S. Census
1920–1940s	Development of machine tools, simple machines that perform manufacturing tasks without human help
1921	Karel Capek coins the term *robot* for his play *R.U.R.*
1926	The first movie about robots, Fritz Lang's *Metropolis*

1940s	Alan Turing founds the field of artificial intelligence
1951	UNIVAC I, the first line of practical computers for commercial use, introduced
1987	Christopher Langton names his research field "artificial life"

GLOSSARY

accelerometer A sensor that detects the rate of change of position in a specific direction.

action A useful motion that a robot makes, like lifting, assembling, or moving something.

actuator The system that makes a robot's arm move, like a motor or an air-powered or fluid-powered system.

AI *See* ARTIFICIAL INTELLIGENCE.

android Robot that looks and acts human and thinks intelligently. No androids exist.

animatronic robot A robot designed to mimic the movements of a particular species and covered with materials that resemble its skin, fur, or feathers.

anthropomorphic Like a human; describing a nonhuman in human terms. For example, using *arm* to describe a robot's manipulator.

artificial intelligence (AI) A computer or robot's ability to think or do work that is just like human thought or action.

artificial life (A-life) A computer program or a computer-driven physical model that behaves like a living organism, mating, reproducing, passing on characteristics, competing, and evolving.

artificial neural network *See* NEURAL NETWORK.

artificial reality *See* CYBERSPACE.

associative memory The human brain's method of storing information by its content and meaning, so that a small clue can bring whole images or ideas to mind.

automaton Self-operating machine in the days before computers; operated by mechanical energy or water power.

autonomous Characteristic of a robot that moves and works entirely on its own.

axis (plural **axes**)   A straight line that an object can rotate around or objects can be placed symmetrically around.

axon   In the human nervous system, a fiber that brings incoming signals to a nerve cell (neuron).

backward chaining   In artificial intelligence, forming a possible answer to a problem, then working backward through a series (chain) of rules to see if they support it.

biomimetics   The design of robots to mimic natural animal designs.

Boolean logic gate   An elementary computer system that can answer yes-no (also called true-false, 1-0, or on-off) questions.

certainty factors   Numbers used in some expert systems to give different grades of importance to different uncertain facts and information.

chance   How likely something is to happen, such as "there is a 50 percent chance of rain today." Also called *probability*.

charge-coupled device (CCD)   A semiconductor that can be charged by both light and electricity and stores the charges sequentially. Used in robotic vision systems.

"Chinese room"   An AI problem in which a person who can't read Chinese must process cards filled with Chinese characters according to a strict set of rules. Is the person doing intelligent work?

CMOS   *See* COMPLEMENTARY METAL OXIDE SEMICONDUCTOR.

complementary metal oxide semiconductor (CMOS)   A simple, low-power computer chip used to process or store data in some portable, battery-operated devices.

computer integrated manufacturing (CIM) or **flexible manufacturing system**   Automated factory system in which computers control most operations, including the use of robots.

continuous path control   Robotic motion in which only the beginning and ending points are set. The controller calculates stopping points in between.

controller   A computer that is in charge of a robot's overall operations.

cortex   Brain area where information is processed and stored and where thought takes place.

crisp   Ordinary numbers and processing, used in contrast with fuzzy numbers and processing. *See* FUZZINESS.

cybernetics   The science of systems, especially the way the body's systems work like mechanical systems that do the same things. An example is the vision system. The term is now obsolete.

cyberspace (artificial reality) A space or world contained within a computer, rather than in the physical world.

cyborg A being that is part human and part robot. No cyborgs exist.

dead reckoning Method of navigation relying solely on the speed and direction of movement.

decentralized parallel processing The linking of individual single-processor computers so that they work on different parts of a problem at the same time.

decision maker Computer program that decides whether a robot should perform an operation.

dedicated Descriptive of a computer that performs just one operation, such as taking in information from a robotic sensor.

deep reasoning Reasoning that analyzes knowledge, experience, or a problem to find out its basic structure.

degree of freedom (DOF) Measure of a manipulator's flexibility. Each joint means one degree of freedom.

dendrite A nervous system fiber that carries outgoing signals from a nerve cell (neuron).

dexterous or **dextrous** Very flexible, like the human hand. In a robot, an arm or hand with many degrees of freedom.

DOF *See* DEGREE OF FREEDOM.

domain Term used in artificial intelligence to describe an expert's specialty, such as medical diagnosis or oil drilling.

end effector A robot's hand—the part that performs a task or holds a tool to perform a task.

exobiology The biology of living things on worlds other than Earth.

exoskeleton The skeleton on the outside of a living thing, such as the natural skeleton of a lobster or as an aid to a physically disabled person.

expert system (knowledge-based system) AI computer program containing an expert's knowledge and method of problem solving.

extravehicular activity (EVA) A term used by the National Aeronautics and Space Administration (NASA) for astronauts' work on the outside of a spaceship or space station. Popularly called a space walk.

first generation expert system An expert system that uses shallow reasoning.

fixed stop Type of robot whose moving manipulator can stop only at preset points.

forward chaining   In artificial intelligence, working through a series (chain) of rules to reach a conclusion or solve a problem.

fully autonomous   *See* AUTONOMOUS.

fuzziness   A method of dealing with uncertainty in which ordinary numbers are expanded into ranges, such as expanding the crisp number "five" to include the range from "four" to "six."

generalization   An intelligent ability to take a specific situation and apply its characteristics to other situations.

graphic-based programming   Method of teaching a robot by designing its motions in a computer graphics program, then transferring them to the actual robot.

grid computing   A large formal linkage of single-processor computers so that they can perform decentralized parallel processing.

gripper   A robotic end effector, such as tongs, suction cup, or magnet, that grasps and holds an object.

halo orbit   A small-diameter orbit around a fixed point in space.

haptics   The formalized field of the study of touch or telepresence. Haptics is from the Greek word for touch.

heuristic reasoning   *See* SHALLOW REASONING.

hierarchy   A group's arrangement of operation by several levels. One level may be more important or complex than another. Or one level may control another level.

hippocampus   Brain area that coordinates memories stored in various parts of the cortex.

humanoid robot   A robot designed to resemble the human body, with a head, two arms, and two legs, and to move like one.

hydraulic   Characteristic of the use of pressurized oil or other fluid for robotic motion.

hypothesis   A possible solution to a problem.

image analysis   Finding the meaning in an image seen by the human eye or a robot's vision system.

inertia   Tendency of an object to remain in a state of rest or of uniform velocity in a straight line, unless an external force acts on it.

intelligent teleoperated   Descriptive of a robot, operated by a human, that has some computer control of the manipulator. Such a robot is called a telerobot.

interface   The part of a computer program that a person sees and uses.

introspection   An intelligent robot's ability to examine its own reasoning methods.

joint Flexible connecting point on a robot, like the "elbow" where two arm links meet or the "wrist" where an arm and hand meet.

joystick A hand-operated omnidirectional control that accurately moves an object in space or in a computer game in two or three directions at once.

just-in-time manufacturing A computerized system for ordering supplies just as they are needed, to eliminate the costs of storage and reduce the time between payment for the supplies and the sale of finished products.

knowledge-based system *See* EXPERT SYSTEM.

Lagrange point 1 (L1) The place where the Earth's gravitational pull is the same as the Sun's.

lead-through programming Method of teaching a robot by moving its arm through the motions. Each starting and stopping point is recorded for playback on the job.

link A rigid part of a robot's manipulator (arm), similar to a human bone.

LISP A computer language, used in artificial intelligence, which processes lists of items.

machine tools Machines that can perform simple factory operations without human help, but are not as flexible and intelligent as robots.

manipulator A robot's arm, which moves and performs a task.

mass The physical measurement of a body's inertia or its ability to accelerate when a force acts on it. In the presence of gravity, such as on Earth, mass is called "weight." In zero gravity or microgravity, such as in space, the term is used instead of "weight."

master-slave manipulator Simple robotic arm that is controlled by a human operator's (master's) hand movements at one end. The arm and hand (slave) perform the same movements at the other end.

memory A person's knowledge and experiences, stored in the brain.

micromechanics Study of the movements of very lightweight, micrometer scale machines within a gravitational field and their difference from the movements of larger machines.

microrobot A robot too small for a human to see with the naked eye, for instance, one that is 0.004 inch (100 micrometers) in diameter.

milling machine A rotary cutter for shaping metal tubes.

mind The thought processes of the human brain.

module A self-contained unit that performs specified operations within a large system.

nanorobots Robots no more than one-thousandth the length of a microrobot.

natural language processing Computer analysis of human language so it can be used in artificial intelligence.

neural network or **neural net** Computer or computer program that solves problems in a brainlike way. It finds the best path that leads from a question to its answer, such as by recognizing patterns.

neuron A brain cell or nerve cell.

node A place where two lines or paths meet. In a neural network, a place where information is transferred from one path to another. Sometimes called a neuron.

nonmonotonic reasoning Reasoning in uncertain conditions.

numerical control Method of programming a machine tool that uses numbers to describe the tool and its movements.

optical computer Computer that uses lasers or other light to process information, instead of using electronic circuits.

pantograph A series of adjustable parallelogram-shaped armatures used as a leg or arm system in some robots.

parallel distributed processing The way the brain processes information. It processes many parts of a problem at the same time and in many different brain locations.

parallel processing Use of several computer chips to work on different parts of a problem at the same time.

pheromones Chemical signals that provide information and trigger subsequent behaviors in members of the same species.

pick-and-place A type of robot that picks up an item and sets it down in another place.

piezoelectric effect The creation of electrical fields and shape distortion in certain crystals when subjected to pressure.

pitch Rotation of an object around its lateral (side to side) axis, such as the nose of an airplane moving up and down.

planner An intelligent computer program that sets goals and priorities for problem solving.

playback *See* RECORD-AND-PLAYBACK ROBOT.

pneumatics Use of pressurized air for robotic motion.

point-to-point control Robot motion system in which the robot can stop only at points it has already been taught to stop at.

probability The chance that something may happen or the degree to which something will happen. For instance, a 50 percent chance that it will rain today.

probe A thin tool used by a surgeon to explore a wound or an incision. Also, an unmanned craft designed to penetrate an unknown region of space and send information back to Earth.

processor A computer.

program A list of computer instructions that tells a robot how to perform a task.

Prolog Computer language used in artificial intelligence to write rules for problem solving. It can handle uncertainty.

real time Time as measured by a clock or your body. Robots must be able to perform tasks or solve problems in real time.

reboost A maneuver that adjusts a spacecraft's altitude to maintain a stable orbit.

record-and-playback robot A robot whose motions are taught and recorded with lead-through programming, then played back on the job.

remotely operated vehicle (ROV) Type of underwater vehicle that is controlled remotely by communication lines (usually tethers) from ships or platforms. Some ROVs are robotic, with arms and specialized hands.

roll Rotation of an object through its long axis, for instance, an inward tilt by an airplane or a bicycle on a curve.

second generation expert system Type of expert system that uses deep reasoning.

semantics The meaning of language.

sensor A human organ that measures touch, sound, light, odor, or taste. A robot part that measures such factors of the robot's environment as light, sound, touch, or temperature.

sensory substitution The brain's ability to evaluate an environmental element with another sense when the primary sense is missing, such as a blind person's substituting touch for vision (the principle of Braille).

sequential processing *See* SERIAL PROCESSING.

serial processing or **sequential processing** Computer processing of one item at a time. The method used by most computers.

servo-controlled Descriptive of a robot controlled by a servomechanism.

servomechanism A device that knows where a robot's joint or arm is in relation to its possible range and feeds the information back to the controller.

shallow reasoning (heuristic reasoning) Reasoning that takes past experience and makes easy-to-use rules out of it.

sol   A Martian "day," one complete rotation of the planet.

somatosensory cortex   Portion of the human brain that finds the meaning of information from the hand's touch sensors.

strain gauge   Instrument used as a robotic force and torque sensor.

stride   The time between a human's or humanoid robot's steps.

submersible   Small submarine that carries a crew and sometimes passengers. Some submersibles are robotic, with arms and specialized hands. The research vessel *Alvin* is an example.

supervised autonomy   Robotic control using computer decision making, but allowing a human to take back control.

synapse   Place in the nervous system where the outgoing signal from one nerve cell (neuron) becomes the incoming signal of another nerve cell.

syntax   The structure of language.

teach pendant   A control box that records the starting and stopping points taught to a robot with lead-through programming.

teleoperated   Characteristic of a robot that is operated by a human, much like a master-slave manipulator.

telepresence   Remote sensing of the environment that gives a human operator the sensation of being in direct contact. A feature of intelligent robotics.

terminator   The line that separates light from dark on a rotating planet or moon.

tether   An electronic line between a human operator and a telerobot that transmits commands, information, and sometimes the robot's electric power.

torque   Twisting.

track   A motion system in which wheels roll inside an endless belt or tread. Used in some mobile robots, it is the same method used on earth-moving machines.

translation   A robotic movement in which two parts stay in the same relative directions. For example, two parts of a sliding joint, which moves like a sliding door.

tribology   The study of the design, friction, and wear of moving parts as they rub against each other.

Turing test   The test to show whether a computer can think, devised by Alan Turing. The computer must perform an intelligent act so that a human expert can't tell whether it was done by a human or a computer.

ultrasonics   Sounds beyond the range of human hearing.

uncertainty   The quality displayed by a situation or question that cannot be understood or answered with yes or no.

visual cortex   Portion of the human brain that processes images that the eyes see.

work path   The straight or curved path a robotic arm follows as it performs a task.

world   The part of the total environment that an individual or a robot can measure with senses or sensors and interpret or understand.

yaw   Rotation of an object around its vertical axis, such as a spacecraft turning while remaining level.

FURTHER READING

Benford, Gregory. *Beyond Human: The New World of Cyborgs and Androids.* New York: TV Books (distributed by HarperCollins), 2001. Companion book to a 2001 PBS documentary.

Bing, Alison, and Erin Conley. *Robot Riots: The Good Guide to Bad Bots.* New York: Dorset Press, 2001. Insiders' view of BattleBots, the battles, and the TV shows that feature them.

Capek, Karel. *R.U.R. (Rossum's Universal Robots),* translated by Paul Selver. New York: Doubleday, Page & Co., 1923. The play that introduced the word *robot.* This edition is interesting because it contains photographs of the first New York performance of the play. This and other editions of the book are available in major public libraries. Library systems that do not have the book can usually obtain a copy through interlibrary loan.

Eckold, David. *Ultimate Robot Kit.* New York: DK, 2001. Suitable for middle school students, the kit contains a book, robot parts made of cardboard and plastic, and a power unit from which users can construct four different robots and then design one of their own.

Grand, Steve. *Creation: Life and How to Make It.* Cambridge, Mass.: Harvard University Press, 2001. The author, a game developer and programmer, defines and makes the case for intelligent artificial life, blending such fields as philosophy, ethics, artificial intelligence, and biology.

Iovine, John. *Robots, Androids and Animatrons.* 2d ed. New York: McGraw-Hill Professional, 2001. Book contains instructions for building robots from both standard parts and home-crafted ones. Robots include: robotic arm, robots with sensing systems, a robot that uses an expert system and a neural network, a robot insect, and a telepresense robot.

Jones, Joseph L., Anita M. Flynn, and Bruce A. Seiger. *Mobile Robots: Inspiration to Implementation.* 2nd ed. Wellesley, Mass.: A.K. Peters Ltd., 1999. Suitable for high school and college students, the book uses photos, diagrams, and text to give a technical overview of robotics and provides instructions for building two robots.

Krasnoff, Barbara. *Robots: Reel to Real.* New York: Arco Publishing, 1982. An entertaining, though out-of-date, overview of robots as depicted in the movies.

Meadhra, Michael. *Lego Mindstorms for Dummies.* Indianapolis, Ind.: Hungry Minds, Inc., 2000. This book provides complete instructions for building 12 Mindstorms robots and includes a CD-ROM containing operating programs.

Papert, Seymour. *Mindstorms: Children, Computers, and Powerful Ideas.* New York: Basic Books, Inc., 1980. Description of the computer language Logo by its major author, along with the philosophy behind it. The book contains a good explanation of Logo, a simple robot language, and how it can be used for education and entertainment. A classic book.

Poe, Edgar Allan. *The Complete Edgar Allan Poe Tales.* New York: Crown Publishers, 1981. "The Man That Was Used Up," on page 192, is the first American work of fiction to depict a forerunner of a cyborg. Poe is often called the father of the detective story.

Ward, Mark. *Virtual Organisms. The Startling World of Artificial Life.* New York: St. Martin's Press, 1999. Book provides a history of the field of artificial life since its beginnings in the late 1940s.

ROBOTICS-ORIENTED
WEBSITES

Military Robots

Office of Naval Research—Supports U.S. Navy and Marine Corps science and technology projects

www.onr.navy.mil/

Search Engines

Find Articles—Search site for magazine articles

www.findarticles.com/   Enter Robotics in search box

Google—Site provides fast and comprehensive searches for text sites

www.google.com   Enter subject name in search box, then click Google Search

Google—Site provides fast and comprehensive searches and displays of images

images.google.com   Enter subject name in search box, then click Google Search

Robotics News

Generation5—Site provides comprehensive coverage of robotics and artificial intelligence

www.generation5.org/   Enter Robotics in search box; also check headlines and links on first page

Technology Review—magazine (published by MIT) site

www.technologyreview.com/ Enter Robotics or other keyword; also check headlines and links on first page

Wired News—Site covers business, culture, politics, and technology

www.wired.com/ Enter Robotics in search box

Yahoo—General news site

dailynews.yahoo.com/ Enter Robotics in search box

Small-Robot Competitions

BattleBots—Official website contains schedule of live and televised battles and other news

www.battlebots.com

BattleBots IQ—Official competition site, with rules and other information

www.battlebotsiq.com/manage.rules.php

FIRST—Official competition site

www.usfirst.org/

RoboCup—Official competition site

www.robocup.org/

San Francisco Robotics Society of America—Official site contains information on West Coast sumo wrestling and other robot competitions; also news and activities

www.robots.org/

Trinity College Fire-fighting Home Robot Contest—Official site

www.trincoll.edu/events/robot Click on information wanted

Space Robotics

Canadian Space Agency—English-language site (also available in French)

www.space.gc.ca/home/index.asp Choose Robotics or other topic from CSA Quick Index

European Space Agency—Official site for the general public

www.esa.int/export/esaHS/   Search for topic

European Space Agency—Official site for present and potential users of the *International Space Station*

www.spaceflight.esa.int/users   Click to enter site, then read information or select topic

NASA—Official site

spaceflight.nasa.gov/   Click on Search, then enter search topic

National Space Development Agency of Japan (NASDA)—Official English-language site

www.nasda.go.jp/index_e.html   Click on a topic; or click on Search, then enter keywords

STORES AND WEBSITES
SELLING ROBOTS, ROBOT KITS,
AND/OR ROBOT BOOKS

Robot Store
Cambridge, MA
San Rafael, CA
www.robotstore.com/

Robot Store
Hong Kong, China
home.hkstar.com/~huip/

SCM International
Capital Federal, Argentina
(Spanish-language site)
www.cybernomo.com/virtual/
2000 Robot Store

Korea
(Korean-language site)
www.robot.co.kr/

Sci-Fi Station
Nevada
www.scifistation.com

Giant Robot
Los Angeles, CA
www.giantrobot.com/

Robot Books.Com
www.robotbooks.com/

INDEX

Italic page numbers indicate illustrations.